We can join the quest for the exceptional.

We were never supposed to be imprisoned in the "ordinary," the daily tedium that limits us to our five senses. We are nonordinary! We are part of an amazing and greater whole, linked to infinite intelligence and power. But as scholar and teacher Nuri Hassumani, PhD, points out in *Infinite Mind*, rather than embrace the wisdom that is available to us, we filter it out. And so, we remain small.

Weaving together stories of extraordinary events and the science of the miraculous, *Infinite Mind* reveals our true potential. Seemingly supernatural abilities like healing, telepathy, remote viewing, manifesting, and interactive dreaming are inherent to us all, because they are part of the conscious universe. When we access this skein of brilliance, we can join the quest for the exceptional.

Every sentence in Nuri's book left me breathless and ready for more, committed to rising above the mundane to dance among the stars.

—Cyndi Dale, author of twenty-seven bestselling
   books about energy medicine and healing

Infinite Mind

A Scientific and Spiritual Exploration:
Building a Bridge Between
Inner and Outer Worlds

Nuri Hassumani, PhD

Infinite Mind Publishing

Design by Meadowlark Publishing Services.

Cover illustration by Tom Stewart.

Published by Infinite Mind Publishing.

Manufactured in the United States of America.

ISBN 978-0-57859-287-9

Published 2019

yourinfinitemind.com

This book is dedicated to my sons,
Daniel and Michael; my three sisters,
Ghazala, Rufi, and Sabrina;
and my mother and father,
Aquila and Anver.

With love and hope, it is also dedicated
to the children of the world,
who are the future—among them,
my granddaughter, Madeline.

Contents

Prologue ix
Acknowledgments xvii
Introduction xix

1   Touched by Infinite Mind 1
2   Intelligence in Nature 25
3   Universal Mind 67
4   Mind Matters—
    How Thoughts Create Reality 115
5   Consciousness—
    The True Nature of Reality 153
6   Peaceful Mind 221

Epilogue: Imaginal Activism—
A Promise to the River 287
Appendix A: Nature of Reality 293
Appendix B: Imaginal Meditation 299
Notes 303

Prologue

This book is contextualized not only in science and spirituality, but also in my own life experiences. Therefore, I believe it would be helpful for you to know a bit about me.

My ancestors are from India. My great-grandfather and his ancestors lived in the state of Gujarat in the district of Kutch, later moving to Bombay, where my father was born and educated. My mother was born in the state of Bihar, in the city of Patna. Her father was a barrister-at-law during British rule.

After World War II, my grandfather traveled to London to argue the case for a free India in the British High Court of Justice. When India gained independence from the British on August 15, 1947, my grandfather's dream of a free and united India—Mahatma Gandhi's dream—did not materialize. The differences between Hindus and Muslims had grown, stoked by the British as well as by the local politicians of the day. The result was the partitioning of India into two independent states: India and Pakistan.

The Muslims of India were forced to move to Pakistan, and the Hindus of Pakistan, to India. This upheaval caused one of the greatest migrations in the history of humankind. Millions of people had to leave behind their homes, ways of life, friends, relatives, personal belongings, livelihoods—everything. There was much violence and bloodshed.

My family moved to Pakistan. I was born and educated there and lived a privileged and comfortable life. Then, when I was nineteen, I moved to the United States by myself, and I have lived here ever since. Though I have fond memories of growing up in Pakistan, my sentiments regarding the partitioning of India are aligned with Gandhi's and my maternal grandfather's vision of a free and undivided India. In the corners of my mind, and in the depths of my soul, this longing will always have a significant presence.

Life in Academia

I have been an educator throughout my professional career, but I didn't set out with that aim. Instead, upon leaving Pakistan, I enrolled in engineering at the University of North Dakota, figuring it would lead to a job with a good salary. I studied hard and ended up earning a Bachelor of Science degree in Industrial Engineering, but the economy was in a slump the year I graduated and finding a job as an engineer was difficult.

It turned out to be a lucky happenstance. I decided to teach algebra and trigonometry to freshmen and sophomores at North Dakota State College of Science—and discovered I had a knack for it. When students who found math particularly tricky exclaimed, "I get it!" and "Oh wow, this is not so hard to understand," I realized teaching was the right career for me, and went on to earn a Master in Education degree from Colorado State University.

I enjoyed teaching, but after a few years I got bored with repeating the elementary rules of basic math; after all, those rules never change. So when I was offered an administrative position with a higher salary and a nice office with picture windows, I couldn't turn it down. Over the next several years I served in senior administrative positions in the Minnesota Technical College system, simultaneously earning a doctorate in education at the University of Minnesota.

I served as the youngest and only immigrant president in the history of the technical college system, and I faced numerous problems: political, structural, and symbolic. I spent many sleepless nights worrying about the complexities of navigating the well-being of students and staff, and the long-term viability of the institution. It was grueling, but during those years I learned a lot about life and work.

The leadership experiences I gained provided me with insights and understandings that proved invaluable when I subsequently started teaching in the MBA program at the University of St. Thomas. During my fourteen years there, I primarily taught courses in leadership and organizational theory and behavior. I decided to integrate seemingly disparate topics into my teaching: new findings in science, and ancient knowledge from yoga and Buddhism. I discovered something that proved pivotal for me: most of the students *loved it.* Thus I realized there is an innate hunger in all of us to better understand who we are and the nature of the universe we live in. This awareness has added to my excitement about sharing the insights I present here.

Moved to Write

This book has emerged as a result of my efforts to live a full, healthy, and happy life, even as I saw the world around me collapse and it brought me to my knees. I will briefly describe the problems over which I despaired, but I think that the way I found peace and joy in the midst of my personal challenges and the chaos of the world will be of greater interest to you. I am sure that when you reflect upon the insights contained in the chapters that follow, you, too, will be able to find peace amid the difficulties and hardships of life.

Several years ago I became extremely concerned about environmental destruction, global warming (as it was known then), violence in all forms (wars, oppression, exploitation), and the general lack of awareness and concern about these issues on the part of the public at large. Big corporations had managed to not only corrupt individual public officials, but they had learned how to affect public policy and create rules and laws that were in their interest at the expense of the public's health and well-being, and their lives in general.

As time went on, corruption, deceit, and wrongdoing continued to escalate in government and business. As I reflected upon the deterioration of American ideals, the erosion of reason, the steep decline of democracy, and the destructive downward spiral for all life on earth, I became despondent. I thought about the eventual and inevitable collapse of our ecosystem and life as we know it. I was unable to come to grips with what I needed to do as a responsible citizen of our planet to stanch the madness.

I started to doubt whether it was even possible to make a course correction. I was deeply saddened, and felt helpless to do anything that would create meaningful change. It was at this extremely difficult time that I began to realize something: no matter what happened to our planet and to humanity at large, I needed to live the remainder of my life with integrity and joy.

As I agonized over my personal problems, the human race, and our planet, I took comfort and found inspiration in the words of many wise men and women who had tried to show us the folly of our destructive ideas and actions. I immersed myself in reading, and I attended lectures, seminars, and conferences so I could learn and grow. Though I found deep understanding and kinship among like-minded souls, my angst and concerns were still getting the best of me.

At a loss for a way to make a difference in the world, I began to turn *inward*. And I discovered through both scientific and spiritual literature that at our core, we are deep love and wisdom—that this is our nature—and that our thoughts are more potent than we usually imagine them to be. I came to this understanding through personal experience, reading, reflection, "coincidences" that occurred in my life, and insights gained during meditation and in dreams. These experiences helped me organize the ideas and understandings I present in this book. When we believe ourselves to be loving, joyous, and prosperous, leading lives of meaning and purpose, we can attain the very intentions we imagine.

Through research and personal experience, I discovered that what we think we lack and the problems and struggles we face are rooted in our egos and our misunderstanding of the nature of reality. Our concept of reality is far too narrow: based upon the perspective of our five senses and the materialistic view of mainstream scientists. And we lack peace because we are not in harmony with our fundamental nature. This creates confusion and fear, which is at the heart of our suffering. In our attempt to feel better, we strive for the attention and acceptance of others, or we stay busy to assuage our discontent.

We fail to realize that our anxiety and suffering have been created and nurtured by our own thoughts and beliefs. Our egos cling dearly to a false sense of who we are and to flawed explanations about life and the universe. To numb our feelings of emptiness and insignificance, we seek thrills and pleasures in lieu of purpose and meaning.

But there is good news: it is possible to achieve internal harmony and peace. And in this book, I aim to present knowledge and practices to help you achieve these too.

I have approached this exploration of science and spirituality like the ancient mariners whose maps were incomplete, relying on intuition and experience to navigate uncharted waters. I know there is much yet to discover, and I will no doubt continue to learn and grow: thus my ideas and suggestions are incomplete. Yet I believe they are developed fully enough to be worthwhile to you.

Furthermore, I have not tried to prove any theory or emphasize any particular philosophical or spiritual perspective. I have instead desired to examine some scientific, philosophical, and spiritual understandings and relate them to my own personal experiences. I believe that my experiences, though unique, are also universal. I am sure you share my desire to lead a happy and purposeful life, and I am sure my trials and tribulations are, in essence, similar to yours.

Connection to the Universe

One of my greatest realizations was that I could not find peace without discovering my connection to the universe. For example, happiness cannot be contextualized into a separate and independent life. Instead, happiness is relational to everything in the universe, as well as to our work, relationships, and understandings about who we are and what our purpose in life is. This is why I offer here details about how the new science describes life and the universe, and similar explanations from the ancient wisdom traditions. I have been amazed by the similarities between new scientific findings and Eastern spiritual insights. As you turn the pages of this book, I think you, too, will be intrigued and perhaps even amazed by them.

My old concept of the universe as a vast emptiness dotted with material planets has changed. I now see a universe that is supremely complex, magical, intelligent—and ultimately incomprehensible. Through this discovery, I have come to marvel at our existence and our potential. I've gained greater respect, admiration, compassion, and amazement for all sentient beings and inanimate objects on earth and throughout the universe.

I am grateful to have attained some measure of peace and a deeper knowing during the process of writing this book. And I am certain that my journey into the inner world will continue and deepen throughout my life. I also believe that no matter how you feel right now, you, too, are in the process of finding greater meaning, joy, and peace. The fact that you now hold this book in your hands tells me that your journey has already begun.

Acknowledgments

Infinite Mind is based upon the hard work and dedicated pursuit of the truth by contemporary scientists, teachers, and philosophers and the ancient wisdom of visionaries, yogis, and gurus. Their ideas and findings appear throughout the book. Words cannot express my deep appreciation and respect for these exceptional men and women.

I thank Sheridan McCarthy of Meadowlark Publishing Services for her helpful editing and suggestions for making this book readable. It was a pleasure to work with her.

Tom Stewart's illustration on the cover captures the essence of this book, and I am grateful for his creative and beautiful work. No one else could have done better.

My friend and neighbor Michael Johnson's tech-support was invaluable. Thank you, Michael!

Writing *Infinite Mind* took several years during which the unconditional support of family and friends kept me going. With fondness and appreciation for his friendship and counsel, I extend a very special thank you to Jim Nelson, author of *Running on Empty*.

Introduction

T he chapters that follow are filled with fascinating information drawn from ancient spiritual traditions and from the cutting-edge sciences of quantum physics, biology, neuroscience, and psychology, among other fields. Together, these two wisdom streams fashion for us a road map for how to live well. As I mentioned in the prologue to this book, over years of research and study I have discovered remarkable similarities between modern scientific knowledge and the understandings of wisdom traditions that have existed for more than five thousand years. My intention here is not to contradict or validate any particular contemporary or ancient perspective. Instead, I aim to whet your curiosity and inspire in you a sense of wonder, and help you arrive at your own understandings of how to live a meaningful, joyous, and healthy life.

In these chapters you will find insights, ideas, and suggestions to help you to see life and the universe in an entirely new way. I want you to comprehend the grand magnificence of your mind and recognize your interconnectedness with the universe. Once you grasp these understandings, you will feel empowered and deeply humbled by the power of your mind and the boundless beauty and wisdom of the universe. You will discover that you are an infinite source of love, beauty, and creativity.

It may be difficult for you to recognize your own grandeur and oneness with the universe at first, but as you become familiar with these ideas and practice the activities I recommend, I am confident you will make marked improvements in how you live, work, and love.

In essence, you will not learn anything new here; instead, you will *remember once again* who you really are. As you commence this journey of transformation, you will experience a greater sense of meaning and purpose. You will know more joy and peace. This will not happen immediately because, after all, this book is only a road map. You are the one who will take the journey to experience the transformation I know is possible for you.

In writing this book over the course of a decade, I had many trials, tribulations, doubts, and fears in spite of my growing knowledge and despite engaging in the practices I present here. It was not easy for me to internalize the findings I've shared with you. While in the midst of my writing and research, I often had serious doubts I would complete the book, and I was frequently disheartened by my lack of progress. Somehow, though, I found the strength to go on. I call this to your attention because, like me, you too will have ups and downs in finding how to live well. But I am confident that transformation is not only possible for you; it is your destiny.

Throughout the book you will discover, from a variety of standpoints, that the nature of reality and of who you are differs from how most of us normally understand these things. We begin in chapter 1 with a discussion of near-death experiences, and find that our physical existence resembles a three-dimensional dream. You will learn that upon death, our awareness shifts to an infinitely wondrous domain, and you'll discover that our minds transcend the limitations of time and space.

In chapter 2 we discuss the intelligence, interconnectedness, and language of plants, the sense of right and wrong among animals, the memory and intelligence of water, and more. All of this paints an intriguing picture of the magnificence of reality and life. The deeper understandings you gain here will prepare you to reflect upon the universal mind.

The discussion in chapter 3 reviews accounts of remote viewing, psychic archeology, the remarkable capacities of savants, the behaviors and personalities of identical twins separated at birth, and the phenomenon of extrasensory perception. Each of these suggests the reality of a universal mind. Knowing that every human being is part of the universal mind expands our understanding of who we are and what we are capable of.

In chapter 4 you will encounter theoretical, experimental, and observational evidence of mind-matter entanglement and the ways in which thoughts create reality. You will see that our notions of reality are flawed, and that when we broaden our perspective, we realize our supreme nature. This awareness helps us understand that we are fully capable of creating meaningful and joyous lives regardless of our circumstances.

We continue to contrast our normal and emerging understandings about life and the universe in chapter 5. To align our thinking with the true nature of reality, we examine consciousness, which is fundamental to the existence of all phenomena, tangible and intangible.

In the last chapter I will help you discover your capacity to use the power of your mind to experience greater peace and love. We will explore ways to experience deep balance between our physical existence and our infinite presence in a "magical" universe. You will see that our ultimate purpose is to co-create our personal and collective realities in order to become the higher beings we were meant to be.

I invite you to turn the page and begin this adventure.

1

Touched by Infinite Mind

In addition to our immediate consciousness,
which is of a thoroughly personal nature...
there exists a second psychic system of a
collective, universal, and impersonal nature
which is identical in all individuals.
—*Carl Jung,* The Collected Works of
　C. G. Jung

Like many people, I have wondered about the big questions: Who am I? Why am I here? What is the universe? Is reality as it appears to be? I believe it is our fundamental nature to think about these questions.

But why couldn't I fully accept the explanation of the origins of life on Earth provided in the theory of evolution? Why did I have doubts about the beginnings of the universe as the Big Bang theory explained them? Why did I feel this way when I considered myself a secular scientific type? I wondered what drove me to spend so much of my time and energy investigating and thinking about these fundamental questions when the answers had already been found in science. What was the source of my curiosity? Why was it so intense? Was there something more than my own thinking that made me question widely accepted science?

Was my soul guiding me to greater awareness? Would discovering who we are, and what this universe is, lead me to a more meaningful, fulfilling, and joyous life? I discovered that it would, and in reading this book, my hope is that you will too.

As I reflected on these questions, I also realized that I had always felt there was more to this world than what I experienced through my five senses. I had been fascinated by nonordinary (or parapsychological—psi) phenomena, but I did not have a scientific or any other framework for understanding such occurrences. Psi phenomena include clairvoyance, telepathy, precognition, telekinesis, near-death experiences (NDEs), many inexplicable coincidences, and certain dreams. My efforts to make sense of my intermittent nonordinary experiences kept the fires of my curiosity stoked as I searched for answers.

Nonordinary phenomena have intrigued human beings since we first walked on this earth. Some humans have embraced these inexplicable phenomena and made them part of their lives and cultures. For example, shamans are considered essential members of their tribes who help sick or ailing tribal members by performing various kinds of healings, as well as informing them about future dangers or the pathways to good fortunes. Other humans shun psi phenomena and consider them witchcraft or the works of Satan. And the orthodox scientific perspective of today emphatically claims there is no evidence of any nonordinary or psi phenomenon.

Yet I experienced nonordinary phenomena myself and found these events intriguing; they raised within me considerable curiosity and wonder. I wanted to know the how and why of my psi experiences, and ended up spending several years searching the scientific and spiritual literature for answers. Today, I have a more spacious sense of these phenomena, culled from quantum physics, quantum biology, psychoneuroimmunology, the science of consciousness, and the basic tenets of yoga, Buddhism, and my personal experiences and understandings. In this chapter I will share some of these with you, as well as the insights I gained as I sought to understand them—for these more expansive understandings based on science, spirituality, and my personal experiences have led me onto the path of finding peace in my daily life.

One such psi event was my near-death experience (NDE), which provided insights but also raised more questions.

My NDE: Seeing Through Closed Eyes

When I was fourteen years old, I fell ill with malaria. It weakened my immune system, and as a result, I also developed cholera and typhoid. I was given powerful antibiotic medication, but I was not getting better. I lost a lot of weight and became so weak that my mother had to carry me to the bathroom. One afternoon our family physician came to see me. Before he left, I heard him speaking in hushed tones with my mom in the hallway next to my bedroom. "We may not be able to save your son," he said.

My mom, holding back tears, came and sat next to me on my bed.

"So what did the doctor say?" I asked her.

She was quiet for a moment and then replied, "He said you will be fine. You will get better."

That was not what I had heard, but I was too weak to argue with her and kept silent. She held my hand for a while, and told me that I should go to sleep. She said she would bring me dinner in a couple of hours.

After she left, it happened! As I lay silently in bed, I started to feel a sense of profound peace, akin to what we experience in meditation but much deeper and more serene than I have ever felt when sitting to meditate. My eyes were closed, but I was aware of everything around me.

Out of nowhere, and for no apparent reason, through my closed eyes I watched as a round, white light appear in front of me. In its presence, I became utterly peaceful. At first, I had no idea what I was seeing or feeling. I just knew I was profoundly attracted to the light and wanted to remain in its blissful presence. *Is this light a representation of God?* I wondered. Soon, perhaps a few seconds later, I instinctively felt that it was.

I was only fourteen years old, and no one had ever told me anything about a white light or an NDE. Seconds later I spoke to the light. "If this means I'm going to die," I told it, "I'm ready to go because I am supremely at peace."

Almost simultaneously, I heard what I can only describe as a silent voice in my heart say, "Are you sure?"

I gently responded, "If I have a choice, I would like to live." And I presented the best argument I could for God to consider: "Besides, I have been a good boy."

The light in whose presence I had experienced supreme peace faded away, and soon after that, I fell asleep.

When I woke up that evening, I was feeling better. My mom came to me and said she would bring me dinner.

"No," I said. "I want to have dinner with all of you at the dining room table."

"Honey, I'm glad you have an appetite again, and that you are feeling better, but you are too weak to walk and sit up at the dinner table while we eat," she said.

"I'll lean on your shoulder, and you can help me walk to the dining room," I insisted.

Reluctantly, she agreed, and I had dinner with my family that evening.

My recovery continued rapidly from that day on. I slept well that night and woke up feeling rested and with greater energy than I had had in months. My appetite returned quickly, and I was gaining strength by the day. In roughly a week after my NDE, I felt strong enough to go back to school. My classmates were happy to see me, and they were curious about my illness and my long absence from school. But I didn't tell them—or anybody else—about my NDE. It was such a profound experience, it was hard to put into words. And besides, I didn't think anyone would believe me. I just said that I had been really sick and was extremely happy to be back in school.

Sports were compulsory at the school I attended, and in about two months I was back on my swim team.

Reverberations

My NDE had a profound effect on me: I lost the fear of death. I also learned that there is much more to life than what we usually experience. We cannot fully understand such phenomena, but we can experience them and gain insights and understandings that transcend logical and rational thought.

Much later, beginning in the fifth decade of my life, I began to explore the literature about the nature of reality, psi, the difference between mind and brain, spontaneous healing, and the power of thoughts and belief. I was curious, intrigued, and motivated to find out whether there were any reasonable answers to such phenomena as NDEs and other nonordinary experiences.

Why did I have an NDE? Why did I heal so quickly after it happened? Why do some people have a near death experience and others don't? And what do we know about people who have had NDEs? I discovered that an NDE is a nonconceptual phenomenon, and that therefore, it cannot be logically understood; it can only be experienced. I also learned that NDEs are not uncommon.

In fact, the occurrence of NDEs in patients who have undergone major heart surgery and flatlined during their operations has been well documented in medicine. In follow-up interviews of patients who revived after their clinical deaths, researchers found many similarities in their experiences, such as the blissful feeling I had during my own near-death experience; speedy and complete recoveries, as also happened to me; and losing the fear of death, or rather, having an understanding that in dying we experience a deeper peace and love.

Knowing the experiences of others didn't have any effect on my memory of my own NDE, but it did make me realize that NDEs are a common and natural phenomenon. I also learned that the extent and nature of NDEs vary considerably among people.

A Neurosurgeon's Journey into the Afterlife

I had traveled to Madison, Wisconsin, a five-hour drive from my home in Andover, Minnesota, to attend a conference called "Research on Near Death and the Experiencing of Dying," sponsored by the Promega Corporation's eleventh annual International Bioethics Forum. There, Dr. Eben Alexander, a neurosurgeon, was going to talk about his near-death experience. Because I, too, had an NDE, I was keen to hear what Alexander had to say. Other speakers at this two-day conference were noted physicians and scientists such as Pim Van Lommel, Penny Sartori, Raymond Moody, Eric Weiss, Stanislav Grof, William Richards, Jeffrey Guss, and Marilyn Schiltz.

However, intuitively I was most interested in Alexander's experience. Perhaps it was because for fifteen years he had been an associate professor of surgery at Harvard with a specialization in neurosurgery, and that he had published over a hundred papers in scientific journals like *Neurosurgery, Clinical Oncology,* and the *International Journal of Radiation Oncology, Biology, Physics.* But it turned out to be much more than that!

As soon as Alexander began his talk, it reverberated in me in a way that made me realize why I had been so keen to hear him speak. I was not merely listening to his remarks and processing them with my logical and rational thinking. I was immersed in his persona, his energy, and the authenticity of his voice. It felt like I was absorbing his thoughts, feelings, and most profound messages at the level of my soul. His message was like cool, clear water that seemed to have come from a magical well in the universe to quench my thirst.

I had sat in the third row from the stage where the speakers gave their talks, and as Alexander walked onto the stage, my excitement grew. I sat up a little straighter in my chair and leaned forward to see him better even though I was already close to the stage. He looked confident, relaxed, and at peace with himself.

Even before he spoke a word, it seemed I had always known what he was about to say. When he was being introduced, he sat comfortably in his chair, and it seemed as if everyone in the audience knew he would share a remarkable story, one that he was eager to tell. There was a hush in the crowd as he began. He spoke in a measured, reflective tone that conveyed not only his intelligence, but the wisdom and authenticity of his message.

Alexander gave a candid, moving, and eloquent account

of his journey into the afterlife. I listened with a combination of admiration and respect, quietly celebrating the fact that a neurosurgeon who had been grounded in materialism and reductionist thinking for all his life had gained the courage to speak like a mystic, while not diminishing or being apologetic or dismissive of his scientific understandings and his professional life.

How could Alexander's talk be anything else? I thought. He had come back from a coma to tell us about his near-death experience and what he had learned from it—it was beyond pure gold!

In her essay "Red Sky in the Morning," Patricia Hampl captures the significance of a personal story, writing, "We do not, after all, simply have experience; we are entrusted with it. We must do something—make something—with it." And when such a story is told, "the writers do not really want to 'tell a story.' They want to tell it *all*—the all of the personal experience, of consciousness itself."[1] Though Hampl's essay is in the context of creative nonfiction, for me, reading it captured the way Alexander spoke that morning.

First, he gave an outline of his seven-day coma, one of the most remarkable, unusual near-death experiences ever documented. It was caused by an *E. coli* bacterial meningitis of his brain. This disease is extremely rare in adults, and less than one in 10 million of the world's population contracts it annually. Alexander told us 90 percent of those who are infected by it suffer rapid neurological decline and then die. He was given aggressive intravenous antibiotic treatment for a week, but his body failed to respond, which put his chances of dying at closer to 100 percent.

As he lay in the intensive care unit of Lynchburg General Hospital with intravenous tubes and a respirator keeping him alive, his cerebral cortex was being devoured by *E. coli*. The cerebral cortex is the largest region of the cerebrum in the mammalian brain and plays a crucial role in memory, attention, perception, cognition, awareness, thought, language, and consciousness. So the assumption was that even if he were to survive—which was highly unlikely—he would be no more than a vegetable. By the seventh day of his coma, his doctors had lost hope for any kind of recovery. The attending physician told Alexander's wife that the family should make preparations for his death. The family began to do so, though not everyone was convinced that he would leave them.

Then, miraculously, soon after this terminal prognosis, Alexander opened his eyes, and his recovery began. The doctors and the entire medical staff at the hospital were shocked, for they had never seen anything like this before, nor had such a case been documented in medical history. Yet as miraculous as his recovery was, it was not the entire story, nor was it as compelling as what Alexander experienced during his coma.

The Girl on the Wings of a Butterfly

Early in his coma, Alexander told us, he was in a dark underground world. He heard loud, pounding sounds that felt eerie and strange. He didn't know why he was there, or how he had come to be there. He did not know how to get out, though he desperately wanted to.

Later, after what seemed like days, months, or eternity, something appeared to Alexander in the darkness: a radiant white golden light. Then he heard the richest, most complex and beautiful music he had ever heard.

He wanted to follow the light and sound and felt that this intention carried him higher, into another world: stunning, vibrant, and brilliantly real. Below him lay a lush green countryside, which seemed Earthlike but wasn't. As his flight upward continued, he saw streams, waterfalls, people, among them children. He heard more music, and laughter. It was real! he told us—more real than anyone could imagine.

During his flight, he realized he wasn't alone. There was someone beside him, a girl wearing a simple outfit in colors of powder blue, indigo, and pastel orange-peach, colors that had the same overwhelming vividness as everything else in his surroundings. As the flight continued Alexander realized that he and the girl were sitting together on top of the wings of a butterfly.

When the girl looked at him, it was with pure love. This was not romantic love or the love of a friend. It was more—much more! He described it as the kind of look that makes you feel that any and every hardship, difficulty, or sorrow you had ever had was worth it if you could arrive at this moment and be seen by those eyes.

Among the many profound things that took place

during his experience, one in particular stood out. Without using words, the girl spoke to him, conveying messages something like:

> *You are loved and cherished, dearly, forever.*
> *You have nothing to fear.*
> *There is nothing you can do wrong.*

As he heard these words, Alexander told us, a profound sense of relief flowed through him like the wind.

Interestingly, as he spoke these words at his presentation, I, too, felt a sense of peace and love. It was as if this message from the girl on the butterfly wings was meant for me, as well as everyone else in the audience.

Universal Lessons

Alexander told us he had learned much about life and the universe during his extraordinary experience in a coma, including these concepts:

> Everything exists in oneness.
> Love is the central force of the universe.
> Life is eternal.
> Past, present, and future do not exist in real time.
> All knowledge is accessible instantly in the infinite
>   domain.
> The physical domain is illusory.

I will discuss these concepts and some others in greater detail in subsequent chapters. For having a better understanding of them is at the heart of finding peace in daily life.

Although while he flew on the wings of a butterfly, he received complete, detailed answers about anything he asked and instantly understood those answers, when he returned to the earthly domain, his brain could not process that information. Still, he knew the answers were inside him. He explained this paradox by using the analogy of a chimpanzee who comes to know all of the Romance languages, calculus, quantum physics, and everything in modern science for a day. But then, when he returns to being a chimp, he has to explain everything he learned to other chimps in chimp language.

Alexander was adopted as a baby. When he was in his fifties, he reconnected with his biological family. That was when he learned that a sister he had never met died when she was a teenager. Four months after he was released from the hospital, his biological family sent him a picture of the sister. He was happy to receive it and pleased to see that they resembled one another. But when he took a closer look, he recognized the face: it was the girl with him on the butterfly wings! Seeing this picture not only helped Alexander emotionally, but it erased any iota of doubt he may have had about what he had seen, felt, and learned during his near-death experience.

I listened in rapt attention to this story, and as his talk came to an end, I remained mesmerized by the man, his experiences, and his deep understandings. It seemed to me that many others in the audience were too. For there

Get this Book!

was a moment of reflective silence and then the audience broke into applause and gave the doctor a standing ovation.

Later I read Alexander's book, *Proof of Heaven: A Neurosurgeon's Journey into the Afterlife,* which was published shortly after his talk in Madison. I reflected upon his coma, his remarkable recovery, the images of waterfalls he described, and the insights he gained while he was brain dead. Why had he traveled on the wings of a butterfly, seen lush green valleys and the brilliant colors of flowers, plants, and trees? Why were they so familiar to him and yet more brilliant, stunning, and captivating than the same images on earth? Would someone who had lived and died in the frozen tundra of Siberia or in the Arabian desert see what he had witnessed? Does infinite mind or the power within speak to us in metaphors to which we can relate? Or, were the images, sounds, feelings, instantaneous wisdom, and oneness he experienced with consciousness literally true? Do we need to remain limited in our binary thinking of true and false, or are there more expansive ways to understand complexity, paradox, wonder, and the mysteries of who we are and what reality is? In adherence to the words of Max Planck, winner of the Nobel Prize in physics in 1918, "It is not the possession of truth, but the success which attends the seeking after it, that enriches the seeker and brings happiness to him."

So, let's continue to probe.

Hugged by a Stranger

On February 2, 2006, two years before Eben Alexander had his NDE, Anita Moorjani, a young woman of Indian descent who grew up in Hong Kong, also had an extraordinarily remarkable near-death experience. The account of her experiences in the afterlife is so similar to Dr. Alexander's that the commonalities further emphasized to me the truth and reality of NDEs in general, and helped me better understand the significance of my own NDE.

For some reason or another, I, too, had a glimpse of a blissful, infinitely loving, and expansive Presence, which ignited my curiosity about nonordinary phenomena, though I didn't realize it when I was a teenager. The memory of my NDE lay dormant in the recesses of my mind until I was in my mid-fifties, when my desire to know more about the nature of reality and nonordinary phenomena was reignited.

During the early days of this intense curiosity, I came across Moorjani and Alexander's books about their NDEs, both published in 2012. My interest in NDEs and the publication of these two groundbreaking books in the same year, containing similar accounts, was an interesting coincidence. I will explore coincidences further later in this book, but it's worth noting here that coincidences are more than they are typically understood to be. Our thoughts, intentions, and actions play a significant role in the coincidences we encounter.

I was a big fan of Dr. Wayne Dyer and had been moved by his presentations on many occasions when he spoke on PBS about finding peace, joy, and love in our lives. I had also been inspired by his numerous well-written and informative books. So when I heard him introduce Anita Moorjani as one of the most genuine and remarkable human beings he had ever met, and framed her NDE as an extraordinarily fascinating account, I was more than keen on finding out what she had to say.

The television setting for Moorjani's interview with Dyer was a park bench with plants and flowers in the background. It was lovely. There was a brief moment of silence as Moorjani took a breath before she spoke. Her voice was strong, her words powerful yet gentle. She conveyed authenticity, truth, and deep knowledge about the love and serenity she felt during her NDE. In her brief interview with Dyer, Moorjani presented a remarkable story, and at the end Dyer informed the audience of her upcoming book.

I anxiously awaited Moorjani's *Dying to Be Me: My Journey from Cancer, to Near Death, to True Healing*. When the book was published, I read it with awe, wonder, amazement, and a growing realization that I was reading profound truths that are not accessible through our five senses, or through orthodox understandings and interpretations about life and the universe based on science. Upon finishing this book, I felt gratitude toward Moorjani for sharing her NDE.

I continued my studies, reading other books that dealt with nonordinary phenomena and the nature of reality.

Then, approximately two years after I read *Dying to Be Me*, I learned Moorjani would be speaking at the Roy Thomson Hall in Toronto, along with several other noted speakers, including Joe Dispenza, Brian Weise, and others whose work I was by then familiar with. But my primary interest was in seeing Moorjani, and I quickly booked a flight to Toronto.

I was among more than twenty-five hundred people attending the event. On the first day, after hearing a few speakers, we broke for lunch. It was a pleasant day and I decided to walk to a nearby restaurant. Several others from the conference went there as well, so the place was filled with lively, energized people. After lunch, as I walked back to the auditorium, Anita Moorjani was on my mind; I knew she was scheduled to speak that afternoon. As I approached the building, I saw a woman walking toward me who looked like her. And much to my delight, it *was* her! My face must have done a very good job of conveying my affectionate admiration for her as I greeted her warmly.

"Oh my God, it *is* you—Anita Moorjani!"

She smiled and acknowledged, "Yes, it is me." Then, instinctively, and without hesitation, we hugged each other.

As we held each other, I was taken aback by the coincidence of having run into her among the throngs of conference goers, and I excitedly said, "I know you well from your book, but I am a complete stranger to you."

She looked at me affectionately and said simply, "It's very nice to meet you."

I felt as if a special gift from somewhere beyond time

and space had just been presented to me on a sidewalk in Toronto. How had this happened? Was it merely a coincidence? As I entered the auditorium after bidding Moorjani goodbye, I didn't think so.

Later that afternoon, when it was her turn to speak, I was thrilled to hear her story. Of course, my excitement was accentuated by the fact I had just met her, talked to her, and even gotten a hug from her. But I believe there were higher forces at play behind how I felt that afternoon.

When Moorjani walked onto the stage, she was instantly greeted with applause of recognition and approval. After making a few introductory comments, she began her talk by telling us that she had grown up in Hong Kong. Her parents were conservative Hindus who had wanted their daughter to grow up to be a proper Hindu Indian housewife. They enrolled her in a British high school where the student body was primarily children of expatriates. There, Anita was not accepted as an equal; she was often picked on by white students, who teased her for being an Indian or for having brown skin. Her classmates were Christians, and they attended church on Sundays, while she was raised in a traditional Hindu family and attended a temple.

Being a young girl, Moorjani was confused about what to believe and how to think, and through those challenging years of isolation and bullying by her classmates, she suffered silently. But she was a good student, and in spite of what she encountered socially, she developed a love of learning and set her sights on college. But her father told her he wouldn't allow it; instead, he insisted she have an arranged marriage.

Contending with parental expectations and what she wanted in life was an internal conflict Moorjani struggled with and remained conflicted about for a long time, but eventually, she got her way with her family and was free to pursue her education and a career. She also found her future husband, a man of Indian ethnicity who was open-minded and encouraged her to pursue her dreams. In 1995, they married, and they enjoyed a healthy, happy life together until 2002—when Moorjani was diagnosed with cancer.

Moorjani emphasized to us that her Hodgkin's lymphoma was partly due to the internal psychological conflicts she had endured during her youth, but that it was primarily because she had become fearful of everything! Contaminated water, plastics, microwave ovens, air pollution, pesticides in food, and the fear of getting cancer. As she put it, "I was afraid to live and afraid to die."

On February 2, 2006, her condition had worsened to the point that she couldn't get out of bed. Her face, neck, eyes, and an arm were swollen like balloons, and her organs were failing. She was taken to the hospital, where, in spite of her failing organs, doctors determined they needed to administer chemotherapy. She entered a coma and, while unconscious, had a near-death experience.

The next day, on February 3, she regained consciousness and her dramatic recovery began. She told the assembled conference goers in some detail what she had experienced during her NDE and what she had learned from it. Despite having already read about her experiences and the resulting understandings about life and the nature of reality she had gained, I found myself riveted.

Moorjani described floating out of her body, to which she couldn't relate, and hearing doctors and nurses talking in other parts of the building about her condition. She was concerned about her husband and her mother and how bad they felt about her. She tried to tell them she was fine, but she couldn't communicate her thoughts and feelings to them. She was aware that her brother was on a flight to Hong Kong and to her bedside.

While Moorjani wanted to be with her family, she was being pulled away from them, as well as from her body, which lay motionless in the ICU. Being body-less felt wonderful, freeing. At the same time, she somehow knew she would be fine when she reentered her body. She felt enormous, unconditional love in the NDE domain, where there was no judgment, only peace and pure bliss. She experienced oneness with the entire universe and felt supreme compassion for all sentient beings. There was no right or wrong in that realm, only beauty, love, and a deep connection with everything and everyone. It was as if *she* was the center of the universe!

I looked around the hall as she spoke and saw tears of joy and amazement in many eyes.

Moorjani's attending physicians were stunned by her recovery, and concluded that the chemotherapy she had been administered a few hours earlier was not the cause of her recovering organs. The lemon-size tumors throughout her body had shrunk dramatically. How could billions of cancer cells be dying, the doctors pondered, even though her organs were failing? They had previously concluded that her open skin lesions would require reconstructive surgery, yet now the lesions were healing dramatically on their own.

While the doctors didn't understand Moorjani's spontaneous remission, *she* knew what had happened. She had been in the realm of pure unconditional love and wisdom: what Eben Alexander referred to as Heaven. This was the place where he too had learned, understood, and gained infinite wisdom and love. Like Alexander, she also knew without any doubt that she would heal completely and that she was returning to her body for a purpose: to share her story with all of us and many, many others.

Similarities Among NDEs

Over the last several decades, modern medicine has inadvertently given us compelling validation of near-death experiences. NDEs were known but rare before the advent of highly effective emergency medical treatments. Since then, advanced practices of resuscitation and life support have saved millions of people who might otherwise have died, resulting in a considerable increase in the number of people who have had NDEs. Not everyone who has been saved from death has had one—but an astonishing number have.

In 1998 Dr. Jeffrey Long, a specialist in radiation oncology, founded the online Near Death Experience Research Foundation (nderf.org). His aim? To learn whether NDEs were real by directly asking those who had experienced them to share their stories. The answer to his question turned out to be a resounding "Yes!" Despite variations, such as in the length of their comas, how they nearly died, and other specific details, the stories of thousands of people—some of which are posted on his website—share many similarities.

One consistently reported experience is people feeling and seeing themselves floating out of, and hovering above, their physical bodies: seeing their bodies on an operating table, or at the scene of an accident, as if from outside. While in that state, they report knowing, seeing, and hearing what is going on, including things that are happening outside the range of ordinary seeing and hearing. After returning to waking awareness, many accurately describe conversations and activities that took place while they were unconscious.

Not everyone who has had a near-death experience comes forward to share their story publicly. Fearing ridicule, feeling they will not be understood, or simply wanting to protect themselves from cynical misinterpretations, they keep their experiences a secret, as I did. Or they share their stories only among a few trusted family and friends.

The current understanding among most scientists and doctors is that NDEs are nothing more than a natural extension of our ability to dream: in other words, mere hallucinations caused by biochemical processes confined to the brain. These assertions, though spoken with conviction, have not been proven but are generally held to be true by the public at large. The layperson is accustomed to believing doctors and scientists, and therefore, their interpretations, rooted in scientific materialism, go unchallenged.

My dreams vary from one night to the next, as do

everyone else's dreams. So to me, the suggestion that similar accounts of NDEs from hundreds of thousands of people living all over the world are merely an extension of our ability to dream sounds more like scientific dogma than an explanation. And it does not explain how people know things that happened around them while they were comatose. How did Moorjani know what the doctors and nurses were talking about in a different part of the hospital while she was "out"? How did Alexander recognize the girl with him on the butterfly wings as his biological sister, whom he had never seen in life? And why, if NDEs are just dream-like hallucinations, does the experience leave people so deeply changed?

Another significant similarity among NDEers is the nature of their memories about the experience. Everyday memories in the local domain are subject to change over time, and memories of the same event witnessed by different people vary in terms of what each individual perceived at the time. In contrast, the memory of an NDE remains rock solid, vivid, and clear over time. The memory of any nonordinary phenomenon, such as clairvoyance or precognition, for example, occurs in the nonlocal domain and thus is fundamentally different in nature from the memory of an event in the local domain.

From my own experience of near death, and from learning about the experiences of Alexander, Moorjani, and hundreds more at nderf.org, I have gleaned a critical lesson: that there is continuity of life, and that after physical death we experience an infinitely greater sense of peace and love. I have also learned that we don't have to die to experience peace, love, joy, and happiness. A regular practice of meditation creates peaceful feelings, as well as a greater sense of compassion, joy, and well-being. In other words, we can get a taste of Heaven without going anywhere at all.

* * *

As I stated earlier, it is not my intent to prove or disprove any scientific or spiritual conclusions; nonetheless, I have tried to understand them. Today, previously held assumptions about the nature of reality are being refuted by scientists and philosophers. New scientific studies point to incongruencies about previously held "truths." And it turns out that mechanistic descriptions of reality do not hold up. Instead, we are discovering a universe of infinite choices and interconnectedness. And, interestingly, these new discoveries are in alignment with the ancient knowledge of sages, saints, yogis, as well of those in our contemporary world who have had near-death experiences.

While mainstream scientists, and the public at large hold on to old conceptions, present-day philosophers have provided congruent and logical arguments that are in alignment with the new discoveries in science. Modern philosophical understandings of the nature of reality and scientific evidence are covered in the chapters that follow, beginning with the topic of intelligence in nature, discussed in chapter 2.

2

Intelligence in Nature

Look deep into nature, you will understand
everything better.
—Albert Einstein

Attack of the Crows   *Unbelieveable!*
smart birds!

W hen I was seventeen years old, we lived on Queens
Road in Karachi. The road had mature trees on
both sides, and many birds made those trees
their home, crows being the most numerous. After school
I hunted birds with my BB gun. It wasn't easy to shoot one
down. It was hard to see the birds in the tall, leafy trees,
and my BB gun was not very powerful; most of the time
the small pellet simply bounced off a branch or twig. When
I had a clear shot at a small bird, I aimed at its head, and

if it was not too far up in the tree, I ended up with a bird for dinner.

The crows stood out among the leaves because of their jet-black feathers, so I started hunting them—not for eating but just to shoot them, for the fun of it. When I managed to kill a crow, the others cawed loudly, as if they were scared and angry. I was too young to know better than to kill crows just for the thrill of seeing one fall from the treetops, and I must have killed a dozen or more crows in that first week of hunting them. Every time I did, the rest of the crows made quite a racket, and the volume of it seemed to increase with every crow I shot. A part of me felt guilty for shooting them, but the excitement I felt when I hit one and watched it fall kept me going.

Then one morning, dressed in my school uniform of a starched white shirt, khaki shorts, maroon tie, and shiny black shoes, I was walking on our long driveway toward the front gate when suddenly from behind me crows started swooping down on me, one after another, attempting to claw my head—hundreds of them in all, cawing loudly and angrily as they dove. Luckily, I had my school briefcase with me. I covered my head with it, crouched down, and ran, and only sustained a few scratches.

As my heart raced, it didn't take long for me to figure out why I was being attacked by a mob of crows: they were letting me know in no uncertain terms that I should not be killing them. As I ran, I decided then and there that I would not kill any more crows.

The next day at the same time, as I was walking on the driveway headed to school, the crows attacked again.

But this time there were fewer of them, and I managed to escape without injury. It happened a third time the next morning, but there were even fewer crows and the attack was less ferocious.

As the days went by and I recalled this fearful experience, I marveled at the crows' intelligence. I wondered if they had communicated with each other about how to let me know very clearly that I should not be killing them. I imagined there might have been various points of view among them: for example, should we hurt the hunter or not? They may have arrived at the decision to attack after discussing several options. Perhaps one option, obviously ruled out, had been to poop on me instead and make a lot of noise to simply scare me. Had the crows reasoned with one another that this might not convey the right message, or that I might get angry and kill even more of them? Perhaps, after proving they could hurt me badly, they had agreed not to wage an all-out war on me by clawing me more severely—at least that was my thinking at the time. Regardless, I still find it amazing that the crows chose to use just the right amount of force, teaching me a lesson about not killing them while showing me how intelligent they were. I didn't think they attacked me simply out of their instinct to survive.

They had figured out the exact time I left in the morning to go to school—when I would be walking alone and without my gun. They had planned a surprise attack from behind so I would be caught completely off guard and startled. They were relentless, swooping down on me one after another so I would not have time to defend myself.

They may or may not have thought about my briefcase and how I could use it to protect my head, but their planned attacks, precisely at the right time for three days in a row, had to have been more than instinct.

I was too embarrassed to tell anyone in my family about these incidents, because I felt I had been wrong to kill the birds in the first place—and because a flock of crows had outwitted me!

Today when I look back at what the crows taught me several decades ago, I realize that intelligence is universal in all forms of life. It makes me wonder if the wisdom of coexistence among all species was better understood by the crows than by me, for they never bothered me again after I stopped shooting them. Did they know about our *oneness* and our symbiotic relationship? Did they want to make me aware of the deep balance that exists in nature? I didn't know it back then, but the lessons the crows taught have been valuable to me, and I'm grateful for them.

Does my interpretation of the crows' behavior have merit? Let's review some recent scientific studies and find out. One showed that when crows feel threatened by humans, they attack them in much the same way they did me. Researchers in Seattle, Washington, who captured crows to brand them and study their behavior reported that afterward, the birds scolded the person who had caught them, cackling and dive-bombing them in mobs of thirty or more. In their five-year study, these same researchers found that crows remember human faces for a long time, even teaching their offspring to identify the specific humans who have threatened them.[1]

The Wisdom of Birds

So yes, crows are smart, but so are many other birds. A recent scientific finding reported by Irene M. Pepperberg, PhD, of the Department of Psychology at Harvard University, showed that parrots don't simply mimic human sounds; they understand the meanings and concepts conveyed by the words and sentences they learn. Her research with an African gray parrot demonstrated that he could understand numbers, shapes, and colors, as well as the concepts of smaller and bigger. Pepperberg noted that this parrot had the intelligence of a five-year-old human child.[2]

Current scientific research provides ample evidence of the intelligence of birds, as well as of all other life forms. Of course, intelligence in all other creatures differs from our own, but remarkable similarities exist. The PBS program *NOVA* ran an excellent series titled "Inside Animal Minds: Who Is the Smartest?" that presented findings on the intelligence of many different animals. I found the information on the intelligence and problem-solving capabilities of New Caledonian crows of particular interest.

Take the case of a crow named Bran, who had been reared from ten days old by a researcher named Lloyd Buck. To demonstrate how smart this bird was, Buck placed a piece of meat in a plastic bottle and crushed the bottle. He also placed a plastic pool filled with water close by, while Bran was perched on his forearm, watching. With a soft, "Go get it" from Buck, the bird hopped off, eager to get at the tasty morsel lodged in the bottle. He couldn't reach it with his beak, and at first he pecked the bottle and kicked it around. Quickly realizing that shaking the bottle would not get the meat out either, he picked up the bottle and dunked it in the pool. When it was approximately a quarter filled, Bran lifted it out of the pool and turned it on its side to pour out the water. Sure enough, along with the water, out came the meat he was after.[3]

Members of the corvid family, which includes ravens, magpies, jackdaws, and jays, are not only good problem solvers, but one of them, the Clark's nutcracker, also has an incredible memory. Research scientist Brett Gibson at the University of New Hampshire observed that this bird stores up to 33,000 pine seeds in caches of up to 5,000 each within a fifteen-mile area for its winter food supply. The nutcracker's remarkable memory allows it to find the seeds, which it had buried in the fall, even after the winter snow arrives. The pine seeds and caches are small and spread over a large area, so the nutcracker's memory has to be exact to find the buried food. As Gibson says, "For us it would be very difficult to remember where we put 33,000 items, but these guys do it really well ..."[4]

Stephan Harrod Buhner, in his book *Plant Intelligence and the Imaginal Realm*, writes that the nutcracker uses various landscape features, such as certain rocks, as markers to triangulate the exact location of the thousands of caches it buries. The process of triangulation requires the bird to establish the exact distances among these markers with only its eyes, and it remembers all the triangulation calculations for more than a year. In addition, the nutcracker remembers the locations of the caches it has already eaten so it can avoid returning to them.[5]

The fascinating memory and intelligence of Clark's nutcracker have been observed in other members of the corvid species and in birds in general. For example, the amazing intelligence of pigeons was documented by Professor Shigeru Watanabe of Keio University, whose study showed that pigeons can be taught to tell the difference between the paintings of Monet and Picasso. Not only could the pigeons distinguish between these artists; they could generalize their stylistic differences. Monet's work is representational (they look realistic) and the pigeons generalized this style to other artists such as Cezanne. The impressionism (not a true representation of the world) of Picasso was also generalized by the pigeons to other similar artists.

The pigeons could even learn the highly subjective human concepts of "good" and "bad." Watanabe used paintings drawn by Tokyo elementary school students to teach the pigeons the difference between good and bad paintings. Artwork that was polymorphous (made up of many different factors) was considered good, and paintings with fewer varied factors were considered bad. Pigeons had no problem discerning good art from bad, and could even generalize these judgments to artwork they had not seen before.

The ability to discern good from bad shows that pigeons are able to understand categories and concepts, and even though pigeons use different pathways to process visual information, there are similarities between the brains of birds and humans. "Bird brains" are capable of learning human concepts such as beauty.[6]

Having learned of these studies and gained a new understanding of the remarkable intelligence of birds, I couldn't help but think that the attack of the crows I experienced as a teenager was indeed an intelligent operation.

Intelligence Takes Myriad Forms

Another segment of *NOVA* showed that monkeys have a sense of fairness, equality, and ethics. How was this demonstrated? A researcher gave two monkeys in separate adjacent cages cucumbers, a food that both monkeys enjoyed eating. After a short while, one monkey was given even tastier grapes while the other one was given only cucumber. Witnessing the monkey eating grapes, the other monkey became very agitated, shaking the cage and slapping the floor to let the attendant know that he was being unfair.

Upon witnessing the monkey's human-like expressions of anger, agitation, and frustration, I was taken aback by the intelligence behind it. The monkey left with only cucumber must have understood that the human attendant was not being fair. He also must have known that his protests could result in him getting some grapes after all. It is one thing to say monkeys are 99 percent genetically similar to us, but in observing the monkey's behavior, I realized just how similar we are to our genetic first cousins.

Concepts of right and wrong, fairness, justice, and equality are not just human ideas and traits. Rather, they exist prominently in monkeys to varying degrees and in other animals as well. Dog owners know firsthand the range of emotions their pets demonstrate. Many dog owners have observed happiness, sadness, guilt, shame, remorse, and the full spectrum of human emotions. The commonality of intelligence and emotions, albeit in varying degrees, exists in all species and is indicative of our shared intelligence and the universality of consciousness.

We have judged intelligence in animals based upon our own intelligence, and therefore either underestimated or overlooked other forms of it in the animal kingdom. For example, an Emory University neuroscientist found that dolphins use their sonar ability to see fish hidden under the sand. They can do this because their auditory nerve is wired to both the temporal lobe and the brain's primary visual region.[7]

The existence of differences in our intelligence from that of dolphins, monkeys, dogs, and birds doesn't mean that one is better than the other. These differences allow each species to function in the environment in which it lives and are therefore useful. At the core of differences in intelligence lie deeper similarities, which become evident when we look more closely and with open hearts and minds. Recognizing the deep similarities in the intelligence of all species allows us to shift from feeling separate to feeling the connectedness, oneness, and peace inherent in all of life. When we dwell on our interconnectedness, we also permit ourselves to feel the bonds of love, respect, and appreciation for all living creatures.

CBS's *60 Minutes* aired a show on October 21, 2018, in which a young woman named Lauren McGough exemplified interconnectedness, appreciation, respect, and love for all of nature. She was particularly captivated by the golden eagles of Mongolia. At the age of seventeen after reading a book on the falconers in that country, Lauren fell in love with the art of catching animals by using eagles. By the age of eighteen, her fascination with eagles, their remarkable interactions with humans, and the stark beauty and vastness of Mongolia motivated her to go there. The falcon hunters she met—all of whom were men—took McGough under their wings and taught her to hunt using golden eagles. Within two weeks she had caught an eagle of her own and begun learning how to train it and hunt with it. As it turned out, she had a knack for communicating with these animals, and today she is considered one of the best falconers in the world. This young woman who "followed her bliss," as Joseph Campbell would say, lived among the nomad falcon hunters for five years. It was heartwarming to see her deep appreciation, love, and respect for the eagles, the hunters' way of life, and all of nature.

Limitations of Awareness: The Robin Puzzle

Once, two robins pecked on the sliding glass doors that open to the deck in my backyard. To these robins, the house inside looked like a safe place to build a nest, but they couldn't figure out how to get through the doors, so they kept pecking at them. Their persistence fascinated me. I sympathized with their inability to understand the solid nature of a transparent door. I talked to them, silently and aloud, in a loving manner, telling them, "You can't come through the solid door. You should go and find another safe place to build your nest." I wonder if the robins heard me.

I know that at a deeper level they did, but their conscious awareness of reality didn't allow them to see the solidity of transparent glass.

So I decided to communicate with them differently; I put a screen door behind the glass. When the robins saw the screen, they stopped pecking, and I watched with relief as they flew away, knowing they would find a more suitable place to build their nest.

The robins' inability to interpret reality reminded me of my own blind spots stemming from the limitations of my five senses. But I also remembered that there are ways to get around some of the problems sensory limitations create. How can we do this? By recognizing our blind spots, listening to the whispers of our intuition, and acting on those whispers. This is not a conscious or rational process. We cannot create magic, though when magic happens, we can be amazed and grateful.

* * *

Throughout the animal kingdom we find that animals have emotions, form relationships, and have the capacity to solve complex problems. Elephants, for example, have enduring memories and mourn their dead. Birds can solve complicated puzzles. Chimpanzees make rudimentary tools. All species inhabit utterly different worlds and come in a wide variety of shapes and sizes, yet exhibit similar forms of intelligence and emotion. What might be the reason for all animals, including us, whether they walk, swim or fly, to have such similarities? Does the answer lie in the existence of universal intelligence or consciousness? We will explore universal intelligence further, but first let's look at the intelligence in plants.

Hugging a Tree

In the early days after my divorce from a marriage of thirty-three years, I had time to think about what had happened and what was yet to come. During those days I was haunted by deep longings to be with my children, and my opportunities to do so were minimal. My older son had started college and was busy with his studies, sports, and his new life away from home. My younger son was involved with his school activities, homework, friends, and the life of a young teenager in general. Rationally, I understood this, but emotionally I couldn't tolerate being away from them.

Remaining emotionally healthy became harder with each passing day. During those days I went for walks in a nearby park and often cried silently, feeling overwhelmed. I so wanted to be with my children and to let them know how much I loved them. Even before the marriage ended, I knew intellectually that being away from them would be difficult. In fact, I had mentally prepared myself to experience grief, remorse, sadness, and loneliness. But I had not anticipated how very difficult it would be. I started seeing a therapist, but that didn't help.

One cold and snowy afternoon I went for a walk in the park and was overcome by grief. I literally fell to my knees, and I didn't want to get up. I sat there holding my head in my hands. I was cold, alone, and depleted, and I didn't know what to do.

I looked up again and saw a tree standing a few feet away. Without much thought, I instinctively walked over to the tree and touched it. At first, I just leaned against it. The support felt good, so I put my arm around its trunk. Almost immediately I felt a surge of energy go through me. I sobbed as I held the tree, and without hesitation, doubt, or really thought of any kind, I asked the tree to help me. "I know you are alive, and it feels good to hold you." I spoke to the tree as if I were talking to a human being—a dear friend.

I put both arms around its trunk and hugged it. A part of me was concerned that someone might see me like this. *What will they think?* But it didn't matter: talking to a tree felt perfectly normal. I held on to the tree for quite a while—it was difficult to let go even though I was getting cold. When I finally left, I promised the tree I would return. I thanked the tree, patted its trunk as I said goodbye, and left in awe of what I had experienced. Amazingly, it held more healing power than my visits to the therapist.

Why had this happened? Why did it feel so good to hug a tree? What made me talk to a tree as if I were speaking to a person? I didn't have any answers, but I felt that what I had learned from hugging a tree was real and important. Some years later, I would have a better understanding of why hugging a tree felt so right.

Intelligent, Social, and Caring Trees

That was the first time I hugged a tree, but it would not be my last. I continued to do it on occasion when I felt moved to do so. Then, a few years after that first experience, I learned that Suzanne Simard, PhD, professor of forest ecology at the University of British Columbia, was scheduled to speak at a conference in Madison, discussing her extensive research on the intelligence and social nature of trees. How could I not attend? Excited, I immediately registered, and I was not disappointed.

Dr. Simard shared her story of becoming a forester, complete with pictures of the forests where she had grown up. From there, she delved into her research. It was fascinating to learn that forest trees are intricately connected to one another through their root systems, that trees function in communities, and that within the forests live elder "grandmother" and "grandfather" trees that communicate their understandings about the environments they live in to their seedlings. These elder trees, she explained, nurture their seedlings by sharing water and carbon reserves with them. There is much more to the story of the life and wisdom of trees, and I will share some of it with you in the next section. (A similar presentation by Dr. Simard in a TED talk can be found at https://www.ted.com/talks/suzanne_simard_how_trees_talk_to_each_other?language=en.)

Relationships Among Trees

It turns out that there is an entire underground world connecting trees with one another, which allows them to communicate in such a way that the forest functions as a single organism. This network represents a form of intelligence that is universal to all life, including inanimate objects. Let's look at this "wood wide network" that Simard discovered in her research, along with other evidence that trees talk to and care for one another.

Simard grew up in Canada, where her grandfather, a horse logger, taught her about the ways of the woods. He cut trees selectively from the forest where he and the rest of his family and grandchildren all grew up. Interacting with the forest was a way of life for young Suzanne.

When she was a little girl, her dog fell into the outhouse at their lake cabin. As she watched her grandfather rescue the poor dog from the muck below, she noticed fungal white mycelium and saw extensive underground tree roots, which sparked her curiosity to learn more and grew into a life-long passion. Her love of trees and forests, along with her concern for the extensive clear-cutting in Canada, inspired her to study forestry.

When Simard first began working as a scientist, she discovered that pine seedlings in the lab shared carbon with each other. This made her wonder if trees also shared carbon in the natural setting of a forest, and she decided to find out. The other scientists in her orbit thought she was crazy to study this, despite seeing her lab results. She could find no funding for such a study, so she had to carry it out, as she put it, "on the cheap."

Simard went to the local Canadian Tires store and purchased plastic bags, shade cloth, duct tape, a paper suit, a timer, and a respirator. From her university she borrowed high-tech equipment, including a Geiger counter, scintillation counter, mass spectrometer microscope, and some syringes, which she needed so she could inject radioactive carbon-13 and carbon-14 as tracers into the small trees.

On her first day in the forest, a mama grizzly bear chased her into her car, but Simard was not put off. The next day when the grizzly was off hunting for berries, she ventured out and started her experiment. She put on her paper suit and respirator and used the syringes to inject tracer isotopes of carbon into saplings: carbon-13 into fir and carbon-14 into birch. Then she covered the little trees with plastic bags so they couldn't share carbon via the air. This, she had hypothesized, would permit her to see two-way tree communication. As she was finishing, the mama grizzly came after her again. This time, holding her syringes over her head, Simard ran to the safety of her car and simply waited for the grizzly to leave. An hour later, which she thought would be sufficient time for photosynthesis to produce carbon in the trees, she went to examine them.

She pulled the plastic bag off the birch and checked it with her Geiger counter. Immediately she heard the crackling sound of the instrument, indicating the presence of the radioactive tracer carbon-13 she had injected into the fir tree. She was thrilled! Next, she went to the fir, pulled off the plastic bag, and ran the Geiger counter. Once again, to her sheer delight, she heard "the most beautiful crackling sound" indicating the presence of carbon-14 sent by the birch. Thus she made her astounding discovery: trees in the forest exchange carbon through their roots. "They communicate and take care of one another!" she told us, beaming with excitement. Her enthusiasm was infectious, and we burst into applause.

Follow-up research conducted by Simard's graduate students further revealed that trees' exchange of carbon varies with the season and with their need for carbon. When the birch needs extra carbon because of its growth cycle, the fir provides it, and when the fir needs more carbon the roles are reversed. "So, trees are interdependent" explained Simard, adding, "They look out for each other's growth and well-being. They are cooperative!" Discovering cooperation in nature shed new light on the dogma of competition and the survival of the fittest. On the contrary, the underground communication network she discovered and refers to as the "wood wide web" is responsible for the massive communication that goes on between trees in the forest. This discovery was published by *Nature* in 1997.[8]

Simard, who has now conducted hundreds of studies, suggests we need to pay attention to the ways in which trees communicate and forests function. Trees communicate not only in the language of carbon, but also through nitrogen, phosphorus, water, and hormones. In other words, in all these ways, they are sharing information.

Simard further revealed to the conference-goers that

underground fungi spread among tree roots and form an extensive network connecting all of the trees in the forest. Where the fungi and roots connect, an exchange of carbon and other nutrients takes place. Fungi depend on the trees for their nourishment, and up to 30 percent of the sugars the trees produce are shared with the fungi. In exchange for this food, the fungi help connect trees in the forest. This currency of food in exchange for connection maintains the entire forest as a single organism.

The underground web of fungi is so dense that there can be hundreds of kilometers of mycelium under a single footstep, all busy connecting many different species of trees and plants.

As all networks do, these underground networks have nodes and links that are the pathways for information exchange. The nodes may be big or small, depending on their age and their connections with other trees. The largest nodes in the network are called *mother trees*, because they nurture their young. Lack of sufficient sunlight under the dense forest prevents seedlings from producing their own nutrients, so the mother trees feed them. A single mother tree can be connected to and nurture hundreds of their seedlings, and they occasionally provide nourishment to other trees' seedlings as well, thereby maintaining the health and vibrancy of the forest as a whole. The fact that trees recognize their own offspring and care for them while also looking out for the best interest of the forest as a whole demonstrates trees' considerable social interaction, emotional bonds, and communal nature.

When mother trees are dying, they send their carbon as

well as defense signals to their seedlings. This exchange of information is equivalent to mothers in the animal kingdom who protect their young and teach them to avoid danger. As Simard put it, "So trees talk!" Such comparisons made by a world-renowned forest scientist were delightfully informative, heartwarming, and uplifting.

As Simard described the social nature of trees, their sophisticated language, their underground communication network, and their intelligence, I couldn't help but recall the peace of mind I experienced when I first hugged a tree, and the manner in which it happened. I wondered, *Could that tree have picked up on my emotions? Did the tree somehow understand how desperately I needed help? Did the tree give me loving energy? Was that why I felt better after touching it and talking to it?* Though none of this made any rational sense, somehow it seemed possible.

Let's delve deeper into the life and language of trees.

The Language of Trees

The trees in the forest are vulnerable to the animals and insects that feed on their leaves. To defend against this, trees use an elaborate language based on scent.

The language of scents is not totally unfamiliar to us. In fact, our own scents influence the behavior of others. Scientists believe pheromones in sweat are decisive factors when we choose the partners with whom we wish to procreate. But in trees the ability to use the language of scents is far more pronounced.

For example, acacia trees in the African savannah

prevent giraffes from eating their leaves by producing toxic substances. They also produce a scent (ethylene gas) that neighboring acacias understand to be a warning that a crisis is at hand, and they respond immediately to the scent by producing toxic substances themselves. However, the giraffes are onto this strategy; they walk upwind so neighboring trees can't pick up the warning scent. In this way they are able to enjoy the tasty acacia leaves, at least for a while.[9]

All plants and trees register pain as soon as some creature starts nibbling on them. (I wonder what my vegetarian friends would think of this.) However, unlike the electrical signals that are transmitted in human tissue in milliseconds, plant signals travel at a much slower speed. But slower electrical signals don't prevent plants or trees from responding adequately in protecting themselves. When an insect starts nibbling on a leaf, the plant recognizes the saliva of that insect and produce pheromones that attract the predators of that particular insect who would eagerly devour it.[10]

Trees and plants can also mount their own defense. Oaks, for example, carry bitter toxic tannins in their bark and leaves that either kill the chewing insects outright or affect the taste so much that the insects won't eat. The defensive compound salicylic acid works in the same way for willows.

Plants and trees also produce a complex assortment of compounds to maintain their health and well-being, including antibiotic and antifungal compounds, anti-inflammatory compounds, analgesics, gums, and resins.[11]

The chemical language of plants and trees and their below-ground network of roots and fungi allow them to thrive amid insects and animals because they have coevolved over millions of years. These symbiotic relationships, with which we coexist, demonstrate a complex and sophisticated deep balance. Intelligence exists among all life forms, and we are all interconnected and interdependent.

Human-Plant Communication

As we have seen, trees and plants communicate with each other, as well as with insects and animals. Incredibly, they also have a sophisticated and deep connection with human beings, and they can even "talk" to us. They communicate with our thoughts and emotions in ways we will look at next. Though the very idea of human and plant communication may seem bizarre in the modern world of science and rationalism, when we consider the evidence with open hearts and minds we discover that it isn't irrational or unscientific at all.

On April 28, 2011, I attended the Annual International Bioethics Forum conference titled "Manifesting the Mind." Participants of this conference included luminaries of the scientific and academic world like Richard Davidson, Wade Davis, Kathleen Harrison, Jeremy Narby, Henry Stapp, and others. I was aware of some of their research and eager to see them in person, and felt fortunate they were presenting in Madison, less than a five-hour drive from my house. I was particularly keen on hearing Jeremy Narby, PhD, speak because of his work with Amazonian Indians who claimed they gained their botanical knowledge by directly communicating with plants.

After a scrumptious buffet breakfast of fruit, eggs, pastries, pancakes, juice, and coffee, I was ready to have my brain fed as well. As it happened, Narby was the first speaker. His credentials were impressive, and his topic intriguing. He had grown up in Canada and Switzerland, studied history at the University of Canterbury in New Zealand, and received his doctorate in anthropology from Stanford University. I sat near the front, eager to learn and determined not to miss a thing this rock star anthropologist had to say.

Narby walked onto the stage with a backpack slung over his left shoulder, wearing khaki pants and impressively playing the part of an adventurous and knowledgeable anthropologist. He began by saying, "Howdy," which made me smile, and I felt an immediate kinship with him. He delivered his presentation with humor, modesty, and clarity.

In 1985 Narby went to Peru to help the Ashaninca Peruvians in the Amazon rain forest save their lands from international banks, which were pouring millions of dollars into the development of land belonging to these tribes. He was there to do an economic, cultural, and political analysis showing that the Ashaninca used their forest resources rationally. The developers, on the other hand, wanted to remove the inhabitants, cut down the trees, and turn the forest into cow pastures.

The Ashaninca took Narby under their wing, and he was

grateful for the opportunity to stay and work with them. In his talk Narby described the rainforest as the epicenter of biodiversity in the world. He emphasized, "There are more species of trees, insects, reptiles, amphibians, birds, and mammals than any other place of similar size; more ants on a single tree then all of the British Isles; more species of birds in a valley than all of North America; more species of trees in a few acres of forest than all of Europe." Narby added, "It's a place where life is most fertile and abundant. You can smell it. There is a musty and funky smell there like in a greenhouse."

He explained that the Ashaninca Indians had names for every plant in the forest, and uses for half of them. They used plants for building materials, dies, food, and medicine. The medicinal plants could heal chronic backache and diarrhea, and accelerate the healing of wounds. Narby was curious about these claims. Back then he didn't really believe plants could heal, but he suffered from chronic backache caused by playing too much tennis in his youth, so he asked a shaman if there was a cure. He was told that drinking a particular plant concoction would cure him. He was also told that drinking it would chill him to the bone for several hours and turn his muscles into rubber, and perhaps make him see images, but on the third day after ingesting it he would be fine. Though he didn't think it would be helpful, Narby drank half a cup of the brew.

Just as he had been informed, his muscles soon started to feel like rubber, he was chilled to the bone for six hours, he saw images, and on the third day he was fine. He has remained pain free ever since. "It couldn't have been psychosomatic," he told us, "because I didn't think it was going to work."

Narby asked the Ashaninca how they knew so much about plants, and their answer was enigmatic. When shamans drink ayahausca, they said, their knowledge of plants comes from the plants themselves, and nature speaks in visions and dreams. It was difficult for Narby to take this explanation seriously. As he emphasized to us, he considered it "an epistemological impossibility."

After several months he found himself in a neighboring village drinking manioc beer, which he said tastes like cold potato soup, with some of the Indians, and he asked them how they had gained knowledge of plants. One man said, "Brother Jeremy, if you want to know the answer to your question you have to drink ayahausca, the 'television of the forest,' adding, "You will see images and you will learn things." Narby had tried LSD in Switzerland a few times, so he accepted the proposition.

A few nights later he found himself on the platform of a house with a shaman and two other people. They were surrounded by the sounds of the forest as they drank the hallucinogen, which, Narby noted, is an extremely bitter substance. As the ayahuasca altered Narby's perception, he found himself surrounded by enormous fluorescent serpents fifteen yards long and a yard high. They appeared in colors hundreds of times more brilliant than ordinary reality, colors far more vivid than he had ever seen before. Soon the serpents spoke to Narby telepathically, and he was told painfully true things about himself. The first thing they said was, "You are just a tiny human being." Looking at them, he could see they were right.

Narby clearly understood that his normal rationalist, humanist perspective about reality had limits. He could see his worldview as bottomless arrogance. He wanted to vomit. He stepped over the serpents, walked to a tree, and leaned against it. About seven yards away he saw an Ashaninca woman in her traditional cotton garb levitating off the ground. He vomited colors. He could see in the dark. After a while he returned to where the other men were sitting. Moments later he felt that he was blasting out of his body miles above the planet, and when the shaman, who had been singing a beautiful melody during Narby's mind-altering experience, changed his tune, Narby landed back in his body. He looked at his hand and saw hundreds of thousands of images of the veins of his hands and those of a leaf, the images flashing in such a way that he saw the two sets of veins were identical. There were so many of these kinds of images that it was impossible for him to remember them all. After some time, he fell asleep and slept soundly.

The next morning when he went to freshen up in the river, he picked up a leaf and lifted it against the morning sun. He could see how similar the leaf's texture and form were to his own hand, and he realized how much like a plant he was. And that he was part of nature. What the shaman had said was true. Narby had ingested ayahausca and learned things. He wondered how, upon his return, his colleagues at the university would view him. He was sure they wouldn't take him seriously. So instead, he wrote his dissertation in the accepted rational framework and earned a PhD in anthropology.

The profound nature of Narby's experience with ayahausca, the elaborate knowledge the Amazonian Indians had about plants, and the paranormal phenomena he experienced over the two years he spent there made him wonder what it all meant. His curiosity led him to an extensive journey of researching, analyzing, and trying to make sense of what he had experienced. He subsequently decided to write a book about what he had learned. In his fascinating lecture Narby outlined the details of *The Cosmic Serpent: DNA and the Origins of Knowledge,* in which he presents evidence of similarities between shamanic knowledge and the origins of DNA.

In 1992 Narby attended the Earth Summit in Rio, where everyone was talking about the ecological knowledge of the indigenous people but no one was talking about its hallucinatory origins. Representatives from all over the world, including advocates, politicians, scientists, and others, ignored the explanations given by the rain forest inhabitants themselves because they didn't fit their preconceived notions about the nature of reality.

As I listened to this part of Narby's story, I knew this type of reaction was not uncommon. Human beings have always been reluctant to accept findings that contradict their beliefs. But during his presentation, my hugging a tree and the peace it brought me were on my mind, and by the time the talk ended I was convinced that humans and plants could communicate.

A Cytogeneticist Communicates with Plants

It may be hard enough to accept that Indians in the Amazon can communicate with plants after consuming ayahuasca, but when a highly regarded Western cytogeneticist claims that some of her knowledge came directly from a plant, what are we to think? The scientist in question was Barbara McClintock. Born in 1902 in Hartford, Connecticut, she received her PhD from Cornell in 1927. In spite of being a woman in what was primarily a man's profession in those days, she became one of the world's most distinguished cytogeneticists. McClintock's work was responsible for some of the most important discoveries about the nature of organisms in their environment since Darwin's studies of evolution.

McClintock discovered that genetic changes in corn were not a product of random chance; they were self-regulated. Because her findings contradicted scientists' deeply held beliefs in those days, she was ostracized by most of her former colleagues and not permitted to work. In *Feeling for the Organism: The Life and Work of Barbara McClintock*, Evelyn Keller quotes McClintock: "It was just a surprise that I couldn't communicate; it was a surprise that I was being ridiculed, or being told that I was really mad … Later on, there were years I couldn't talk to anybody about this and I wasn't invited to give seminars either."[12] McClintock reported that, while he was visiting her, a well-known geneticist of the time said, "Now I don't want to hear a thing about what you are doing. It may be interesting, but I understand it is kind of mad."[13]

Like the Amazonian Indians, McClintock had learned by directly communicating with a plant. Here is how she describes the way she discovered how DNA functions in corn: "Well, you know, when I look at a cell, I get down in the cell and look around..."[14] She adds, "the more I worked with (chromosomes) the bigger and bigger (they) got, and when I was really working with them I wasn't outside, I was down there. I was part of the system. I was right down there with them, and everything got so big. I even was able to see the internal parts of the chromosomes—actually everything was there. It surprises me because I actually felt as if I was right down there and these were my friends."[15] McClintock elaborates: "As you look at these things they become part of you. And you forget yourself. The main thing about it is you forget yourself.... "You let the material tell you where to go, and it tells you at every step what the next step has to be ..."[16]

McClintock's "transposition" in corn was considered so ridiculous that the scientific community rejected her work for more than thirty years. But decades later in 1983, when biologists through their own research determined "transposition" was an indisputable phenomenon, Barbara McClintock was awarded the Nobel Prize in Medicine, presented to her by King Carl Gustaf of Sweden.

So, what does it mean for human beings to learn from plants? How is this possible? Are there any answers in biology, botany, cosmology, quantum physics, or the philosophies of the ancient wisdom traditions that provide explanations for these "magical" phenomena? If such explanations exist, what are they? In both quantum physics and spirituality, they do exist, and we will discuss them in greater detail further on. For now, a brief explanation will suffice.

The principle of "entanglement" in quantum physics and the concept of "oneness" in the *Yoga Sutras of Patanjali* both provide a framework for understanding human and plant communication. Both claim that all of nature is fundamentally connected, exists in energetic form, and communicates with everything else in nature. This communication takes place at the nonlocal or subconscious level and can be accessed in different ways, commonly while dreaming or during deep meditation.

Experimental Evidence of Interactions Between Plants and Humans

Cleve Backster did fascinating experiments in the mid-sixties about the ability of plants to communicate with human thoughts and emotions, as well as with other life forms. In 2003 he wrote a book about his work entitled *Primary Perception: Biocommunication with Plants, Living Foods, and Human Cells.* Let's review how he carried out some of the experiments he chronicles in the book and take a look at his findings.

Prior to his work with plants, Backster worked for the CIA, where his responsibilities included using the polygraph (lie detector) to screen applicants for employment as key CIA personnel. Later, while working with other organizations, he continued to consolidate, refine, and expand upon existing polygraph techniques. Along the way he developed the Backster Zone Comparison Technique and the first numerical system for evaluating polygraph charts. By this time, he was recognized as an expert in his field.

Backster's fascinating work with plants started serendipitously on February 2, 1966, around seven o'clock in the morning at his lab in New York City, while he was taking a coffee break. His actual work had nothing to do with plants, but he was watering two plants that his secretary had purchased to have something green in the lab. As he watered them, he wondered if it would be possible for him to measure the rate at which water rose from the root area to the leaf. The dracaena plant in particular, with its long trunk and leaves, caught his attention, and he decided to attach the polygraph's electrode sensors to a leaf of the plant.

In humans the polygraph works on the detection of variation in sweat on the skin, known as the galvanic skin response (GSR). When a person being examined is asked, for example, "Did you fire the fatal shot?" a true response does not cause greater perspiration, but a lie does. Lying elicits a tiny amount of sweat, creating greater electrical conductivity. The increased flow of electricity is recorded on the polygraph chart.

After hooking up the leaf with the electrode, Backster decided to check whether a threat to the plant would elicit a human-like response. *Burning the plant would be a real threat*, he thought while standing fifteen feet away from the dracaena. At the instant the thought entered his mind, the polygraph recording pen moved rapidly to the top of the chart! No words had been spoken, the plant hadn't been touched, and the match hadn't been lit. The mere intention of burning the leaf caused a dramatic excitation in the plant—the plant understood Backster's intentions telepathically! This remarkable event profoundly affected Backster, and over the next several years he conducted many other experiments on plants so as to understand them better.

His subsequent experiments showed that plants could distinguish real human intentions of harm from those that were merely pretense, and that plants developed attunement with their caretakers just as pets do. They demonstrated instantaneous telepathic abilities with human thoughts and feelings from a distance. Furthermore, Backster's research showed that plants had concerns and compassion for all life forms. These included bacteria, other plants, and higher life forms such as brine shrimp.

Backster was keen on publishing his findings, so he consulted a few scientists to help him design an experiment that would meet the protocols and rigor of scientific research. He was a good engineer, so he built a fully automated experiment to record the reactions of plants upon the killing of brine shrimp. Four polygraphs were involved in the experiment, with three hooked to different plants and one as a control for comparison. Brine shrimp in a separate room were to be dumped in boiling hot water by a mechanical "cup dumping device" at a random time. The exact time when the shrimp were dumped in boiling water was automatically recorded, and a ten-minute time delay was built into the device to allow Backster to leave the lab and drive at least an avenue away from his building. He did this because he had learned that when he was in the lab, the plants he was studying remained attuned to him, and therefore did not react to other stimuli.

Having taken these precautions, Backster started the experiment and left the building. Amazingly, the moment the shrimp were dumped in the boiling water, all three plants exhibited noticeable reactions that were detected on the polygraph charts, indicating their concern or compassion for other life forms. A control polygraph, not hooked to a plant, showed a flat line, as anticipated. A report of this experiment was published in the *International Journal of Parapsychology.*[17]

Backster also left the polygraphs hooked to the plants to detect unforeseen reactions when he was away from his lab. One evening he and a colleague decided to go watch a football game at a sports bar about twenty miles away. Upon their return, they saw several significant reactions on the polygraph. At first Backster was puzzled about what could have caused these reactions, but as he thought about the evening, he realized that they coincided with touchdowns during the game as he and his colleagues watched it—and their excited cheering and celebrating. The plant had telepathically picked up on their excitement, from twenty miles away, at the instant it occurred.[18]

How did this happen? Was there any scientific explanation for this seemingly magical communication? Backster later learned that, according to quantum physics, the effects of "entanglement" are instantaneous and occur without expenditure of energy. This and other experiments Backster conducted were full of surprises and led him to new understandings. Nature, it seemed, was keen on revealing some of its secrets.

Backster's experiments were reported in various newspapers and magazines, creating considerable interest among the public at large. Television networks invited him to their popular late-night shows, hosted by Johnny Carson, Merv Griffin, Art Linkletter, and David Frost. On one such show, Frost inquired whether a philodendron plant was male or female. Backster playfully suggested that Frost should go over and lift up a leaf and take a peek. Before Frost even reached the plant, the large meter being used to display the plant's GSR activity showed a wild reaction, evoking applause and laughter from the audience.

Backster was invited to speak about his work at more than thirty-five colleges and universities in the United States, and in June of 1974 he appeared before the ninety-third US Congress to testify about his findings. Yet while the public and the press had considerable interest in Backster's findings, the scientific community ignored his research.[19] However, empirical evidence cannot simply be pushed aside and ignored forever. Some of today's best biologists and botanists have published hundreds of papers about the unique intelligence of plants. These recent discoveries give credence to Backster's experiments, and further research on biocommunication is bound to follow.

As we have seen, intelligence, memory, language, problem solving, emotions, telepathic communication, and social structures are all part of plant life. Now let's review another aspect of plant intelligence. When we consider longevity as an attribute of intelligence, plants demonstrate they are certainly more intelligent than humans. For example, the Great Basin bristlecone pine of California lives for more than five thousand years, and the Pando tree of Colorado, also known as the trembling giant, has a root system that is more than eighty thousand years old and is among the oldest known living organisms.[20]

Stefano Mancuso is professor at the University of Florence in Italy and is the world's leading authority in the field of plant neurobiology. He has published more than 250 papers, and the cumulative findings of his research document that plants are intelligent, have memories, and can solve problems. In his book *The Revolutionary Genius of Plants: A New Understanding of Plant Intelligence and Behavior*, he writes, "After years spent investigating the many aspects of plant intelligence, I have become consistently surprised and fascinated by plants' clear capacity for memory." He explains that plants don't have a central brain; rather, the whole of the plant has the ability to learn, understand, and react successfully to new and trying situations. In his book, which was translated into English in 2018, Mancuso states, "It is impossible to learn without memory, and the ability to learn is one of the requirements of intelligence. Living things are generally capable of learning from experience, and plants are no exception to this principle; they respond in ever more appropriate ways when known problems recur throughout their existence. This could not happen without memory, the ability to store, somewhere in the organism, the relevant information needed to overcome those specific obstacles."[21]

Mancuso is right. Current orthodox scientific research has not been able to explain how intelligence in plants exists without their having a brain. However, the notion that a brain is necessary for intelligence is not based in fact; rather, it's a belief. As we will see later, intelligence and memory reside in the nonlocal domain. The existence of universal consciousness, which contains all of knowledge and is the source of intelligence, was described in Samkhya, one of the oldest philosophical systems of ancient India.

In the West, reference to the concept of nonlocal universal consciousness has become increasingly common. For more than a century philosophers, psychologists, biologists, physicists, cosmologists, and medical doctors have written books and published papers that affirm the existence of universal mind outside the domain of individual brains. Their views often reflect the philosophy of metaphysical idealism, which states that consciousness is the source of all known and unknown phenomena in the universe, and therefore everything has "intelligence" and "memory." We will discuss metaphysical idealism a bit later in this book.

Ralph Waldo Emerson, in his essay "The Over-Soul," writes: "There is one mind common to all individual men. Every man is an inlet to the same and to all of the same. He that is once admitted to the right of reason is made a freeman of the whole estate. What Plato has thought, he may think; what a saint has felt, he may feel; what at any time has befallen any man, he can understand. Who hath access to this universal mind is a party to all that is or can be done, for this is the only and sovereign agent." Emerson continues to assert that wisdom and knowledge emanate from universal consciousness: "Within man is the soul of the whole; the wise silence; the universal beauty, to which every part and particle is equally related, the eternal ONE. And this deep power in which we exist and whose beatitude is all accessible to us, is not only self-sufficing and perfect in every hour, but the act of seeing and the thing seen, the seer and the spectacle, the subject and the object are one."[22]

The Intelligence of Water

I remember the joyful surprise I experienced when I unexpectedly felt a sense of calmness while driving to Duluth, Minnesota. My girlfriend at the time and I had been in the car for approximately three hours and, being in no hurry, decided to stop along the way. Duluth is a picturesque town of approximately eighty-five thousand on the banks of Lake Superior.

When we were approximately fifteen or twenty miles from the lake, I started feeling a deeper sense of ease than I had felt when we started our journey, even though when we pulled out of the driveway, I felt rested, having slept well the night before. This new feeling as we approached Duluth had elements of peace and joy and an air of excitement, and seemed to have come from nowhere. Why was I feeling so good? Where did this feeling come from?

I didn't dwell on these questions and thought it must be because we would soon get a chance to get out of the car, walk along Lake Superior, and enjoy the cool, bright afternoon. Once in Duluth we strolled along the shoreline, appreciating the flowers, plants, and trees along the boardwalk. The memory of my experience a few short miles from the city stayed with me.

During those days I was teaching in the Graduate School of Business at the University of St. Thomas in the Twin Cities. To create meaningful discussions with my MBA students, I sometimes asked questions that were not directly related to the subject I was teaching: leadership. One day I asked, "Why do you think we are attracted to bodies of water like lakes, rivers, and oceans?"

Their responses included "Because, lakes look beautiful"; "I love being in the lake and just relaxing on its shore"; "Feeling the ocean breeze and walking on sandy beaches is calming"; and "The sounds of a flowing river are engaging."

I acknowledged that all those answers were good but added that our attraction may also be because our bodies are 70 percent water. Most students resonated with this observation, agreeing with my suggestion as if remembering something from long ago, and some wondered why they hadn't thought of this themselves. Their appreciation for a lake's beauty, the mesmerizing sounds of ocean waves, the way river water gushes among rocks, or the way the mist of a waterfall feels seemed to reflect the deep, primordial connection we have with water.

I began my discussion on "leadership" by saying that we often miss seeing answers about life and work because we don't contextualize our thinking in a larger framework, and that paying closer attention to understanding who we are often allows us to see things differently.

Though I had experienced deepening peace as we

approached a lakeshore town some years earlier, I didn't really know why. Had I intuitively picked up on the deep peace and "intelligence" of Lake Superior? Today, having read the scientific research about the intelligence of water and our connection to it, I understand why I or anyone else could find peace near water. Let's review this ingenious and fascinating research, which reveals that water interacts with our thoughts and feelings, and that it has memory, intelligence, and the capacity to heal. These remarkable qualities of water and our oneness with it are integral to my understanding that everything in the universe is a manifestation of consciousness.

Based on a series of recent scientific studies conducted to determine whether water responds to human thoughts and intentions, Dawson Church in his book *Mind to Matter* documents that not only does water respond to human thoughts and feelings, but it remembers, and it has the capacity to heal when it has been treated by a healer.

He describes the pioneering research of Bernard Grad of McGill University, who examined the effects of healing energy on plants. When water was treated by a healer for thirty minutes and then used on barley seeds to determine their growth rate, more of the seeds germinated and the resulting plants grew taller. Their chlorophyll content increased, and the quality of leaf growth was significantly enhanced. Other researchers have also documented significant improvements in plant growth or seed germination after healers treated the plants.[23]

While these research studies cannot be understood or explained in the context of mainstream botany, biology, chemistry, or physics, they are nonetheless significant and extremely important! These and many other studies point to the existence of a conscious universe in which everything is aware, interconnected, and intelligent.

Another study, in which water was examined to determine if it underwent *structural* changes when it was treated by therapeutic touch practitioners, demonstrated such changes in the water. The water molecule (H_2O) has two hydrogen atoms connected to one oxygen atom. The angle of the bond between them can be measured, just as you can partially open a hinge and measure the angle it forms. The angle of the molecular bond of normal water is 104.5°. After forty-five-minute therapeutic touch sessions, the water showed statistically significant changes in its absorption of infrared light, which demonstrated that the bonding angle between the oxygen and hydrogen atoms was altered by contact with the healing field. Other researchers have also found alterations to the molecular structure of water after contact with the healer.[24]

These physical changes in the structure of water grabbed my attention, much in the same way I have been awestruck by every new mind-bending piece of research pointing out that the "reality" we experience every day isn't the way it actually is. I am heartened and inspired by scientists who maintain their intellectual integrity and continue to patiently and meticulously probe into evidence even when it turns out to be contrary to their own understandings and beliefs, which have been contextualized in materialism and reductionist thinking.

Researchers at the Chinese Academy of Sciences

conducted ten experiments with qigong master Dr. Xin Yan, who significantly altered the molecular structure of water. In the first experiment he stood near the water. In the other nine, he was at various distances, between seven kilometers and nineteen hundred kilometers away, while allowing a control sample to remain unchanged.[25] Distant healing seems magical, even unbelievable, but in the context of entanglement it simply validates a fundamental principle of quantum physics.

These studies demonstrated that water has the unique quality of being receptive to healing energies. And, as Professor Masaru Emoto's research shows, water also reacts to classical music and loving and compassionate thoughts by forming exquisitely beautiful patterns in water crystals. In contrast, when water is subjected to the heavy metal style of loud music or angry thoughts, the patterns formed in the water crystals appear jagged and nonsymmetrical.[26] It is remarkable that not only does water react to the fields created by human thoughts, but it also maintains a "memory" of these energetic healing properties.

In *Water and Its Memory: New Astonishing Insights in Water Research*, Bernd Kröplin and Regine Henschel document a series of experiments performed at the Institute for Statics and Dynamics of Aerospace Structures at the University of Stuttgart, Germany, demonstrating that each person has a unique "signature" in the form of patterns that appear in drops of water.

These observations were made when a group of participants filled a hypodermic syringe with water and squeezed droplets onto a microscope slide. After the drops of water were allowed to dry, Kröplin and his team took photographs of them. To their astonishment, they found that drops of water squeezed by an individual had a unique and similar pattern, while drops squeezed by other participants demonstrated their own different and unique patterns, much like the differences in fingerprints. It seemed as if the water samples' passage through the energy field of a person had created their own unique patterns, which were recorded and maintained in the water droplet.[27]

As scientific studies have shown, intelligence in animals, plants, and water, is remarkably similar to human intelligence. These similarities have profound implications about who we are, and the interconnected wholeness of everything in nature and the universe. While it is true that there is intelligence in nature, it would be more accurate to state that nature *is* intelligence.

Next we will explore the remarkable intelligence of humans and the *universal mind*.

3

Universal Mind

Even scholars of audacious spirit and fine instinct can be obstructed in their interpretation of facts by philosophical prejudices.
—*Albert Einstein*

Every man takes the limits of his own field of vision for the limits of the world.
—*Arthur Schopenhauer*

In this chapter we will look at the evidence of a universal mind or consciousness, and the interconnected nature of reality. Accounts of a universal mind are found in the ancient wisdom traditions, and today there is scientific evidence for it as well. We will review accounts of remote viewing; psychic archeology; the remarkable capacities of savants; behaviors and personalities of identical twins separated at birth; and extrasensory perception (ESP) phenomena. All of these suggest the reality of a universal mind.

Plato's Cave

In Plato's *The Republic,* Socrates addresses the relationship between appearances, reality, and knowledge with an allegory of an underground cave with its mouth open toward the light of a burning bush. Within the cave are people who have lived their entire lives chained to the walls of the cave with their backs to the entrance, able to see only the wall in front of them and the flickering shadows of the world outside. The cave dwellers equate these shadows with reality, naming them, talking about them, and even linking sounds from outside the cave with movements on the wall. Truth and reality for these cave dwellers are the shades and shadows on the wall, and so it is with the shadows cast by the burning bush.

Then one day an inhabitant breaks loose from the chains and sets himself free. He ventures out of the cave and for the first time sees the burning bush outside. He is thrilled to also see the sun, sky, trees, flowers, birds, and animals, and he realizes that the shadows he had witnessed for a lifetime were just reflections of a more complex reality. He understands that the knowledge and perceptions of his fellow cave dwellers are distorted and flawed.

If he were to return to the cave, he would not be able to live in accordance with his old beliefs. He would find it impossible to accept being chained again, and would pity the plight of the cave dwellers. And if he shared his new knowledge with them, he would probably be ridiculed for his views. For the cave dwellers, the images on the walls would remain more meaningful than a world they had never seen, and they would interpret his descriptions of the outside world as detailing a dangerous place best to be avoided. Thus, far from enlightening the captives, new knowledge that stems from a more complete experience of reality could actually lead them to tighten the grip on their familiar interpretations of the real world.[1]

The existence of a universal mind was first presented in ancient India in the *Upanishads* and later in the *Yoga Sutras* of Patanjali. In the twentieth and twenty-first centuries, new scientific understandings have emerged that are in alignment with the descriptions of Eastern philosophers, yogis, and mystics, but they have not been accepted readily by mainstream scientists or the general public. In this chapter we will review some "extraordinary" phenomena to suggest the existence of a universal mind or consciousness as the bedrock of all known and unknown phenomena of the cosmos. It may be difficult for those of us schooled in Western thought to let go of familiar understandings and beliefs, but a clearer understanding of reality allows the shadows on the wall to appear as what they are.

Let's begin by reviewing the phenomenon of extrasensory perception (ESP). Most of us, at one time or another, have had different types of ESP experiences in the form of hunches, intuitions, or dreams. These inklings may startle or surprise us, and make us wonder how or why they happen, but we normally don't dwell on them; instead, we move on to the activities and responsibilities of our lives. But ESP experiences are not "extraordinary," even though we don't have access to them as readily as we do to our five senses. Moreover, as we will see, they point to the existence of a universal mind.

My Brush with ESP

In August 1967 at the age of nineteen, I moved to Wahpeton, North Dakota, to study engineering, leaving behind everyone I had loved and grown up with in Karachi, Pakistan.

I arrived a week before classes started for the fall semester at North Dakota State College of Science. After unpacking my suitcase, which took about five minutes, I sat on my bed with nothing to do. I did not know anyone, and the campus seemed empty. Feeling lonely, I left my dorm room in McMahon Hall to see if I might run into some students. I had grown up in a city of 15 million, so the sights, sounds, and energy of a big city were in my DNA, and in spite of being on an empty campus, I fully expected to run into other people.

I was taken aback by the ghostly silence of the desolate campus, which was nestled in a residential neighborhood. It was dusk, and the fading light made the grassy oval-shaped area in the middle of campus, lined with pine and maple trees, feel strange and eerie. At a time like this, thoughts of home and family can be comforting, but instead I missed the people and the life I had left behind. I was scared. *How am I going to make it through the next four years if I'm feeling this way already?* I wondered. I wanted to let my parents know I had landed safely, and had reached the school, but back then the only way to contact them was through letters, so I made a mental note to locate the post office as soon as classes began.

As I walked among the buildings on campus, I wondered where my classes would be held. Despite knowing I was coming to a small town in the Midwest, I was disheartened at how small Wahpeton was. I was unprepared for the too-quiet reality of the place, and I felt like I had made a big mistake. Reluctantly, I accepted that my conceptions of the United States, which were largely influenced by movies depicting the life and vibrancy of big cities and of well-to-do middle-class families, needed considerable revision. It was starting to get dark, and I was feeling helpless, scared, confused, and utterly alone.

I turned around and made my way back to McMahon Hall. Near the front entrance of the dorm was the student lounge. As I stood outside the lounge looking through its glass doors, I saw a coffee table and chairs dimly lit by directional lights. The rest of the room was completely dark. Much to my relief, I noticed the silhouettes of four young men sitting around a table. I walked into the room and asked if I might join them, and one of them said, "Sure." I sat down, feeling self-conscious and uneasy. I soon learned that these guys were on the college football team and had arrived a week earlier for practice.

As I listened to their conversation, a thought suddenly entered my mind that the next football player who spoke would say, "I custom-combined this summer." I was surprised and delighted when the very next sentence the man spoke was exactly what I had heard in my mind. But I was also puzzled—I had never heard the expression "custom-combined" before and didn't know what it meant. From the context of his comments, I thought he was referring to some kind of farm equipment for picking crops.

Now I was keen to hear more from this guy, wondering how long my thought-reading powers would last. Initially, knowing what I was about to hear felt normal. Simultaneously, it was amusing to know, without any conscious effort, what he would say. This went on for at least two or three minutes. But then, by the time the football player had finished telling his story, I had gone from feeling amused and amazed to feeling anxious about what I was experiencing. And as soon as I became anxious, my mind-reading ability evaporated.

Perhaps in an effort to dispel my anxiety—or maybe for some other reason—I wondered what was behind me in the dark. Instantly, in my mind I saw a set of shelves with speakers, a turntable, and a stereo. As I examined this mental image further, I saw that on the bottom shelf was a stack of LP records. At the front of the stack was an album with a picture of a woman's face. Curious to find out if what I had imagined was real, I got up, turned around, and walked in darkness toward the shelves that had appeared in my mind.

Just then another image, of a light switch on the wall, popped into my head. When I got close to the wall, which I could not see in the dark but could sense, I reached out, and to my surprise, I felt a real switch in exactly the same spot as it had appeared in my imagination. I turned it on.

Directional lighting fell softly upon a real set of shelves—the exact replica of the mental image I had seen. Again, I was amazed! On the bottom shelf was a stack of LP records, and the one on top had the same woman's face on the cover that I had seen moments earlier in my mind.

At the time, all this made no sense to me. How could I have known what the football player was going to say before he said it? In a dark room, how and why could I see images in my mind that turned out to be real when the lights were on? Why had any of this happened? Was there any way to explain this experience?

The Miraculous Mind

My experiences were similar to the findings in, *The Reality of ESP: A Physicist's Proof of Psychic Abilities* by Russell Targ. This book documents a variety of experiments and findings that demonstrate the human ability to see objects and activities at a distance.

Russell Targ was born in Chicago in the 1930s with poor eyesight, and as an adult he was considered legally blind; it is interesting that someone who couldn't see ended up showing the world how to see at a distance. As a young physicist, Targ developed lasers that were more sophisticated and powerful than the ones available at the time. His work earned him the reputation of being an exceptionally bright and talented scientist.

In 1972 his friend Dr. Hal Puthoff, who was also a laser physicist, started conducting ESP experiments at the Stanford Research Institute (SRI) in Palo Alto,

California. With the help of gifted psychics, over the course of twenty-three years, Targ and Puthoff conducted research and made some remarkable findings for the CIA, NSA, NASA, and the military. Under strict scientific protocol, the double-blind studies they conducted documented that human beings have innate capacities to see objects, events, and phenomena at a distance.

Some of the most fascinating remote viewing experiments at SRI were carried out by Ingo Swann, an artist living in New York City, and Pat Price, a former police commissioner and vice-mayor of Burbank, California. Descriptions of the sites these two men observed through remote viewing were based only on the longitudes and latitudes of the sites provided to them while they sat at a desk, shielded from electromagnetic waves in a Faraday cage. Some of the sites observed at a distance included a secret NSA listening post in Virginia; Mount Hekla in Iceland; the French-owned Kerguelen Island in the South Indian Ocean; details of an atomic bomb manufacturing facility six thousand miles away in Semipalatinsk, Soviet Russia; and the rings around Jupiter—before NASA's space probe confirmed their existence.[2]

Let's review some of these remote viewing experiments, the results of which were published in prestigious journals such as *Nature, The Proceedings of the Institute of Electronic and Electrical Engineers (IEEE),* and *Frontiers of Time: Retrocausation Experiment and Theory,* published by the American Institute of Physics. These research findings have been replicated worldwide.[3]

Ingo Swann's Psychic Abilities

As I mentioned earlier, to evaluate the accuracy of the information obtained by SRI's psychic experiments, the CIA provided Targ and his team with longitude and latitude coordinates for a target site. As he often did, Swann would close his eyes and begin describing what he visualized. An artist, he found it easy to sketch, so he drew what he saw in his mind's eye. These drawings proved to be remarkably accurate. During the experiment Swann's verbal accounts were also recorded:

> There seems to be some sort of mounds or rolling earth. There is a city to the North. I can see taller buildings and some smog. This seems to be a strange place, somewhat like the lawns you would find around a military base. But I get the feeling that there are some old bunkers around. It may be a covered reservoir. There must be a flag pole, a highway nearby, and a river to the far east. There is something strange about this place. Something underground. But I'm not sure.[4]

Later it was found that not only were the drawings correct, but on a map Swann drew, even the distances, flag pole, underground bunkers, and directions were accurate. The selected site turned out to be the secret National Security Agency listening post at the navy's Sugar Grove, West Virginia, facility. Not only had Swann identified the site, but he had also sensed "something strange," signifying the secrecy of the NSA listening post 2,800 miles away.

Swan coined the term "remote viewing" for his ability to see objects at a distance. However, as we will see in the descriptions of other sites he observed remotely, it would also be appropriate to describe his psychic abilities as remote *sensing*.[5]

To further determine Swann's psychic abilities, which interested the CIA, Targ and his colleagues conducted more experiments. Coordinates for Mount Hekla, an active volcano in Iceland, were given to Swann, who was unaware of the actual location. Within a few seconds of learning the coordinates, he expressed feelings of vertigo, sickness, and being cold and described a sense of being at a great height and above a fiery furnace. He said, "I am over the ocean. I think there's a volcano to the southwest."[6]

Swann accurately described the location of the site, but vertigo, sickness, and being cold and scared are all feelings one might experience in actuality while smelling noxious fumes emanating from a fiery volcano. How could Swann feel this way while sitting comfortably in the SRI lab thousands of miles away? How should we interpret such an experience? In reality, beyond the limitations of our senses, in the universal mind everything exists in oneness and at the *present* moment. Therefore, it is possible to be anywhere in the universe instantly. We will discuss this issue throughout the book, but for now let us continue to explore Swan's remarkable psychic abilities.

In another experiment to further document Swann's psychic abilities, the CIA provided coordinates of the French Kerguelen Island in the South Indian Ocean. At the time of this trial, the island was a French and Soviet meteorological station for radar mapping of upper atmospheric research. Without anyone at SRI knowing what the site was except the coordinates that were provided, Swann almost immediately visualized the site and began describing it:

> My initial response is that it's an island . . . maybe [there's] a mountain sticking through the cloud cover. There's something like a radar antenna . . . a round disc. There are some buildings very mathematically laid out. To the southwest there is a little airstrip. It's very cold.[7]

Next he drew a map of what he saw in his mind. The drawings correctly showed an island with many bays and inlets and a large mountain to the west. It took SRI staff two years to confirm the airstrip Swann had drawn, and the accuracy of his drawings put an end to the concerns of doubters that Swann had "memorized the globe."

Buoyed by accurate descriptions of previous experiments, Targ and his team decided to test Swann's ability to describe something that was not even on the planet, a place that was not mapped and about which little could be known or memorized. Swann wanted to see planet Jupiter, and the team at SRI was keen to determine whether human remote sensing ability extended that far.[8]

The "Jupiter Probe," as it was dubbed by Swann and colleagues, took place in 1973 at SRI. The research team had three goals: (1) to determine the distance at which remote sensing was possible; (2) to document the time it took before impressions were made; and (3) to compare the impressions with scientific information, which would be published by NASA based on the Pioneer 10 spacecraft's "flybys," scheduled for four months later, as well as a Pioneer 11 voyage in 1974, and the later Voyager 1 and 2 probes of 1979.

The following is a transcription of Swann's "Jupiter Probe" of April 27, 1973:

6:03:25:
There's a planet with stripes.

6:04:13: I hope it's Jupiter. I think that it must have an extremely large hydrogen mantle. If a space probe made contact with that, it would be maybe 80,000–120,000 miles out from the planet surface.

6:06: So I'm approaching it on the tangent where I can see it's a half-moon, in other words, half-lit/half-dark. If I move around to the lit side, it's distinctly yellow toward the right. (Hal: Which direction did you have to move?)

6:06:20: Very high in the atmosphere there are crystals . . . they glitter. Maybe the stripes are like bands of crystals, maybe like rings of Saturn, though not far out like that. Very close within the atmosphere. I bet you they'll reflect radio probes. Is that possible if you had a cloud of crystals that were assaulted by different radio waves? (Hal: That's right.)

6:08:00: Now I'll go down through. It feels really good there (laughs). I said that before, didn't I? Inside those cloud layers, those crystal layers, they look beautiful from the outside. From the inside they look like rolling gas clouds—eerie yellow light, rainbows.[9]

Astronomers in 1973 did not know whether Jupiter had rings. However, when NASA published pictures of Jupiter taken by the Voyager and Pioneer spacecraft, Swann's description of the rings around the planet was shown to be accurate. The experiment also showed that remote viewing is not constrained by distance and that information about distant objects can be accessed instantly.

In Targ's book *The Reality of ESP*, Swann's sketches are shown side by side with actual pictures of Jupiter and its rings. The similarities are compelling evidence for ESP, a must-see for anyone interested in enriching their understanding of the remarkable capacities of human intelligence and the universal mind.

The story of Jupiter's rings was published in newspapers and magazines all over the world based on the photographs. Today, a Google search shows hundreds of published reports on the rings that confirm Swann's ESP findings. But even though Swann was the first to discover them, he is rarely mentioned; only the photographs the scientists took are included as proof. In any case, what's even more remarkable about Swann's remote viewing of Jupiter is that it was extremely difficult for the spacecraft to take these pictures. The following description on NASA's website explains this difficulty:

> Jupiter's faint rings were first discovered by the Voyager 1 spacecraft in 1979, when it looked back at Jupiter and towards the Sun. They are so faint and tenuous, they are only visible when viewed from behind Jupiter and are lit by the Sun, or directly viewed in the infrared where they faintly glow. Unlikely Saturn's icy rings full of large icy and rock chunks, they are composed of small dust particles.[10]

It is not uncommon for the mainstream scientific community to disregard ESP-based evidence even when it proves factual, but ignoring ESP doesn't invalidate it. In fact, organizations like NASA, the NSA, and the CIA relied on the ESP capabilities of SRI, and for over two decades provided the program with more than $20 million to carry out psychic surveillance activities. Let's examine further the top secret work SRI conducted for the CIA.

Psychic Patrick Price

Patrick H. Price was also a gifted psychic who assisted Targ in his remote viewing experiments. According to Price, when he was a police commissioner, he tracked downed suspects using his psychic powers of remote visualization to catch the crooks. Sitting with a dispatcher in the police station listening to reports of criminal activity, he would scan the city psychically and locate the whereabouts of the suspect, and a squad car would be sent to pick up the perpetrator.[11]

Price's remote viewing capabilities made him a perfect fit for the experiments SRI was conducting. In one top secret experiment, the CIA gave Price the longitude and latitude of a specific site. Sitting in a chair in the lab, he drew sketches and described in remarkable detail what turned out to be a secret Soviet atom bomb laboratory in Semipalatinsk, Siberia. Price leaned back and closed his eyes, as he normally did during psychic remote viewing experiments.

After a few moments he began to describe his mental images:

> I am in the sunshine lying on top of a three-story building in some kind of R&D complex. The sun feels good. Some kind of giant gantry crane just rolled over my body. It's going back and forth. . . . This is the biggest damn crane I've ever seen. . . . It runs on a track, and it has wheels on both sides of the building. It has four wheels on each side of the building. I have to draw this."[12]

Price's sketches of the buildings, gas cylinders, rails, and pipes, as well as a detailed drawing of a gantry, when compared to actual satellite pictures taken later, were remarkably accurate and led the CIA to become interested in knowing what was going on inside the building.

So the next day Price was instructed to focus his attention on the interior of the building, a space that was unknown to the CIA or anyone else in the US government at that time. Price began by describing a large interior room:

> There's a lot of activity. They're trying to make a giant steel sphere. It looks like it's going to be about sixty feet in diameter. They are making it out of "gores" and trying to weld them together. But it is not going well, because the metal is so thick.... The gores looked like sections of an orange peel.[13]

The sphere turned out to be a containment vessel for a particle-beam weapon to shoot down US satellites on reconnaissance missions over Soviet territory. Their existence was verified three years later, in 1977, by US satellite pictures and data. Again, Price's description turned out to be remarkably accurate. He also determined that the Soviets were having trouble welding the spheres at high temperatures, which caused warping in the metal, and that they were interested in finding a lower-temperature welding material. Three years later, even this technical detail was confirmed.[14]

Price died in 1975, and left the world to contemplate the infinite capacities of the human mind. Targ described his friend as an even-tempered "man among men" and recalled fondly the time when Price's young secretary asked if he could follow her psychically into the ladies' room. His reply: "If I can focus my mind on any place on the planet, why would I want to follow you into the ladies' room?"[15]

Targ and his team of psychic viewers performed countless amazing tasks for the government. They found a downed Russian bomber in Africa, located a kidnapped American general in Italy, described the health of American hostages in Iran, and located the whereabouts of Patricia Hearst's kidnappers.[16] These and other accounts of remote viewing were published in Targ's books, as well as some he coauthored with others.

Targ's work helped me understand that my anomalous experiences were not abnormal or absurd; rather, they reflected an inherent human ability. As I mentioned earlier, universal consciousness is the source of everything, and therefore all human capacities are part of an omnipotent and omniscient universal mind. Gifted psychics can access the universal mind at will, and therefore can see objects at a distance, or sense any past, present, and future phenomenon. But almost all of us have ESP experiences once in a while, which occur unexpectedly without deliberate intent. Others claim they have never experienced anything extraordinary because it's not possible, or they simply deny the experience of ESP.

An important piece of the puzzle in understanding ESP and the nature of reality is explained in Targ's *The End of Suffering: Fearless Living in Troubled Times*. In this book, which he coauthored with J. J. Hurtak, he explains the four-part *tetralemma* logic of Nagarjuna, the second-century Indian genius. Nagarjuna's system of logic allows us to go beyond the limitations of our everyday thought processes and binary language. We will discuss this system of logic in chapter 5, which will help us better understand our nonlocal universal mind, the nature of reality, and ESP. For now, it is sufficient to know that all of reality as it is experienced, interpreted, and understood by humans is in the binary logic of true or false. Furthermore, this two-part, Aristotelian binary logic reflects the limitations of our five senses. Nagarjuna's four-part logic includes binary logic with two additional nonconceptual ways to interpret nonordinary phenomena, consciousness, and the universal mind.[17]

The remote viewing experiments we have discussed demonstrate that our psychic capacities enable us to explore an important part of our true nature. The scientific evidence of our psychic abilities and its implications are described in Targ's book *Miracles of Mind: Exploring Nonlocal Consciousness and Spiritual Healing*. He emphasizes that learning to use our nonlocal minds, as in remote viewing, is important because it gives us access to the wider world. Through this personal experience, we understand firsthand that consciousness has no boundaries. We feel empowered by our connection to the infinite, inspired to reach our highest potential and gain an expansive sense of our ultimate purpose.[18]

* * *

In October of 2014, I attended the Science and Non Duality (SAND) conference in San Jose, California, where I had the opportunity to attend Targ's lecture called "Scientific and Spiritual Implications of Psychic Abilities." It was heartwarming to see him in the eighth decade of his life, carefully walking up the stairs to the stage for his presentation. Though he could hardly see, he had learned to view objects at a distance from his colleagues at SRI who could see objects, landscapes, and even planets millions of miles away. The reason humans can see, sense, and know about distant objects or any phenomena, he explained, was that everyone and everything is in oneness in the nonlocal universal mind or consciousness. Targ ended his talk with a smile and quipped, "There are no secrets anywhere."

Finding a Lost Harp

Russell Targ and his coworkers were excellent at remote viewing, but they were by no means the only ones. Let's look at the story behind a stolen harp, how it was found, and how it led an acclaimed university professor on a journey to discover answers to the exceptional and "magical" powers of the human mind and the nonlocal universal consciousness.

In 1991, Elizabeth Lloyd Mayer, PhD, taught in the psychology department of the University of California at Berkeley and at the University Medical Center in San Francisco. Her research interest was about female development, and she had a successful psychoanalysis practice. She belonged to numerous professional associations and was busy with committee work, attending international meetings, functioning on editorial boards, and lecturing all over the country. Mayer was a busy academician who enjoyed a meaningful career and being a mother.[19]

Mayer's eleven-year-old daughter, Meg, had played

the harp since the age of six. One day after she played at a Christmas concert, the instrument was stolen. This particular harp was not a classical pedal harp but a smaller and extremely valuable one built by a master harp maker. A two-month search to find it included contacting the police and instrument dealers across the country, advertising in the American Harp Society newsletter, and a CBS TV news story—but nothing worked.

One day a close friend told Mayer that if she really wanted to find the harp she should contact a dowser, because, according her friend, "The really good ones can locate lost objects as well." Though Mayer didn't believe anyone could find an object with a forked stick, she was desperate. Nothing else had worked, and Meg didn't like playing rented harps. So Mayer called the president of the American Society of Dowsers, Harold McCoy, in Fayetteville, Arkansas. Over the long-distance call, Mayer inquired whether McCoy could help locate a harp in Oakland, California. He replied cheerfully, in a heavy Arkansas accent, "Give me a second. I'll tell you if it is still in Oakland." He paused and then said, "Well, it's still there," and asked Mayer to send him a street map of the city.

Mayer remained doubtful, but sent him one anyway. Two days later McCoy called and said, "Well I got that harp located. It's in the second house on the right on D_____ Street, just off of L_____ Avenue."[20]

Though skeptical about this information, Mayer figured she had nothing to lose, so she drove to the location and got the street address. Next she called the police and told them that she had gotten a tip about the harp's whereabouts.

That information wasn't good enough to issue a search warrant, however: case closed.

But Mayer couldn't let go, and she posted flyers in a two-block area around the house offering a reward for its return. She was embarrassed to be acting on the assertion of a dowser, and told only a couple of her close friends about her search.

Three days later a man called her and said his neighbor had showed him a harp he had recently acquired, and that it matched the description on the flyer. He also mentioned that his neighbor could neither play the harp nor afford to buy one, and that he suspected he had stolen it.

Three weeks passed and the same man called to say he had the harp, and that a teenage boy would return it to her at 10:00 p.m. in a Safeway parking lot. Sure enough, when she went to meet him the young man was there, and he gave the harp to Mayer. Half an hour later as Mayer pulled into her driveway, she was at a loss for how to make sense of what she had just experienced.

The mystery proved the catalyst for an extensive inquiry that led Mayer into the world of anomalous phenomena. She met hundreds of people from all walks of life who began opening up to her about their personal ESP experiences, and she learned that the reasons for their previous silence about these experiences were similar: to avoid being ridiculed, considered crazy, or accused of lying. She also found well-researched information about ESP, and learned that these studies had been systematically rejected by mainstream scientists.[21]

She studied the subject of ESP for years, and in March

2007 published a book titled *Extraordinary Knowing: Science, Skepticism, and the Inexplicable Powers of the Human Mind*. Reflecting on the extensive body of research on anomalous phenomena conducted by some of the best scientists, she wrote:

> Yet as I delved more deeply, what most impressed me was the significant bank of well-conducted, scientifically impeccable research that imposes enormous questions on anyone interested in making sense of the world from a Western scientific point of view. I began to wonder, why had so much of this excellent research been overlooked, its conclusions dismissed?[22]

As a psychologist, Mayer wondered why our culture is so fearful of anomalous experiences. Could this be why so much good research in the United States hadn't been given more attention? She wanted to know the nature of the discomfort and the conflicts that underlie it. And she spent a great deal of time and energy developing an understanding of how these could be resolved.

Mayer was not just speaking for herself. She was expressing concerns she had in common with several groups of people: scientists conducting ESP research; medical doctors who had witnessed miraculous or spontaneous healing in their patients; those who'd had anomalous experiences themselves; and quantum physicists who understood the nature of reality differently from the explanations found in scientific materialism.

I, too, had experienced the discomfort Mayer described with some of my friends and acquaintances. Others were intrigued by my ESP experience, and some shared their own inexplicable experiences with me. However, the answers to how ESP works were not readily available to any of us.

Mayer found that in trying to understand anomalous phenomena, we must first acknowledge their reality, and then realize that they reflect human capacities that cannot be understood within the confines of scientific materialism. But the necessary expansion and openness required for scientific materialism to seek to understand such phenomena weren't going to happen easily.[23]

Thomas Kuhn, in his book *The Structure of Scientific Revolutions*, explains that the established ways of thinking in science become dominant and that any new observations must fit the basic assumptions of the dominant scientific perspective. The gradual assertion of new ideas, he suggests, doesn't change the existing paradigm. Instead, a *paradigm shift* must occur for revolutionary ideas to gain enough force to turn the old dominant understandings upside down.[24]

Mayer remained grounded in the rational world, while being aware that experiences like finding the harp required a different way of interpreting the nature of reality. She worked diligently for fourteen years seeking answers, and wrote, "As human beings, might we be capable of a connectedness with other people and every other aspect of our material world so profound that it breaks all the rules of nature as we know it? If so, it's a connectedness so radical as to be practically inconceivable."[25] Mayer was right about the nature of reality being "inconceivable," for it appears to be infinitely and wondrously complex.

Elizabeth Lloyd Mayer died before her book was published, but she managed to bring attention to the existing knowledge about the inexplicable powers of the human mind. Though skepticism about anomalous phenomena continues in mainstream science, the door to *extraordinary knowing*, which helped her find the harp, had been opened.

Psychic Archeology

Finding hidden treasures through psychic means—as a dowser sitting by the phone in Arkansas located a harp in a house in California—has a long and interesting history. Stephan A. Schwartz, a scientist, futurist, author, and adventurer, has been involved in conducting archeological findings using remote sensing techniques for more than three decades and has documented the history of psychic archeology.

Schwartz's credentials are impeccable. He has authored more than 130 technical reports and papers, twenty academic book chapters, and four books. He was the recipient of the 2017 Parapsychological Association Award for Outstanding Contributions, and was selected by *OOOM Magazine* (published in Germany) as one of the 100 Most Inspiring People in the World.[26]

Schwartz, in his book *The Secret Vaults of Time: Psychic Archeology and the Quest for Man's Beginnings*, documents several fascinating accounts in which ESP was used to locate and reconstruct archeological sites. According to

Schwartz, retired British Major General James Scott Elliot was the world's first dowsing archeologist. Elliot didn't like being called a psychic, however. "I have no interest in the psychic whatsoever. I don't even use the word. Just call me what I am. A practical dowser, that's all."[27] In 1956 he retired from the British Army, and with "nothing really mentally satisfying to do," delved into dowsing and archeology. He learned the basics of dowsing from a local man, and to get better at it, for the next six months he walked along the fields of Dumfriesshire County, on the English-Scottish border, with a metal rod. His neighbors were bemused at the sight of a major general who had led combat troops in World War II in Africa and Italy waving a forked metallic contraption in the fields.

Elliot's efforts to become proficient at dowsing paid off; he became very good at it. At first he helped a few of his neighbors who thought there could be archeological sites hidden beneath their properties. These forays into dowsing were tested by excavation, and proved to be highly accurate.

However, he didn't stop there. One day while working on a large property, he decided that instead of walking the entire premises, he would attempt to describe what was hidden underneath by simply making a detailed map of the property and gaining the required information while observing the map. He used a string and a pointed plumb that he guided over the map to sense the site's details, while sitting comfortably in his office. This technique came to be known as "map dowsing."

Elliot's psychic archeological findings in England included a fifteenth-century grain-drying kiln, a Roman encampment, a Bronze Age crematorium, and a five-thousand-year-old possible precursor to the fabled Stonehenge. All of these were unknown until he pinpointed their location using nothing more than a little metal rod or a swinging pendulum.[28]

How was it possible for Elliot to sense and feel the details of underground objects and sites not known of for hundreds or thousands of years? Was this information dependent on the use of a metal stick or a pendulum? Considering there are psychic archeologists who don't use either, it seems that external props for psychic archeology are useful for some, while others need only information like the longitude and latitude of a site to access details about it. The ability to see remotely, sense hidden or distant objects, and describe the activities that took place at those sites is an innate human capacity of accessing the nonlocal universal mind.

Canadian Psychic Archeology

J. Norman Emerson was a senior professor of anthropology at the University of Toronto, founding vice president of the Canadian Archaeological Association, and a teacher of Canada's anthropologists and archaeologists. He trained many Canadian professionals in those fields in the 1940s and 50s at some point in their careers. His former students included University of Toronto's anthropology faculty members, professors at other universities, and most government archeologists. Because of his work and long tenure in academia, many considered him the father of Canadian archeology.[29]

In March of 1973, speaking at the annual meeting of the Canadian Archeological Association, Emerson said this:

> It is my conviction that I have received knowledge about archaeological artifacts and archaeological sites from a psychic informant who relates this information to me without any evidence of the conscious use of reasoning....
>
> ... By means of the intuitive and parapsychological a whole new vista of man and his past stands ready to be grasped. As an anthropologist and as an archaeologist trained in these fields, it makes sense to me to seize the opportunity to pursue and study the data thus provided. This should take first priority.[30]

If anyone else had made such a statement in front of mainstream archeologists and anthropologists, crowd reaction would have been very different from how Emerson's remarks were received. While some missed what he said, and others were confused by his remarks, most of the group seemed to understand the implications of what they had heard from their esteemed colleague. But why would a scientist, close to the end of an illustrious career, depart from thirty years of practicing accepted methods of scientific research?

Many of his colleagues may have thought Emerson was

simply getting old and losing his grip on reality. However, Emerson had sought the help of a psychic because of his unyielding commitment and desire to know the past as fully and accurately as possible. While traditional archeological methods reveal sites, dates, and artifacts, these methods cannot produce conclusive understandings about the lives of the men, women, and children of the past. Emerson was interested in greater details about the day-to-day life, work, relationships, aspirations, and beliefs of the people who lived in those days, which psychic archeology could provide.

But why would a rational, intelligent, and seasoned scientist think that psychically obtained information could be trusted? How had he arrived at this point of view? Emerson would answer in these words: "I have to confess that after thirty years of work … if you had asked me, I'd have had to say of those questions there was no way I could even have attempted to answer them. Today, however, I would reply that, yes, it may well be possible to do so—with the help of psychic persons."[31]

Emerson's interest in psychic archeology was triggered by a series of coincidences. In the 1960s his wife, Ann, got interested in the paranormal after reading a biography of Edgar Cayce, the best-known and most highly regarded psychic of the twentieth century. His fascinating work and life prompted Ann to join an Edgar Cayce Study Group, where she met a woman named Lottie McMullen and the two became friends. Lottie told Ann about her husband's psychic abilities, and as time went on the Emerson and McMullen families developed a friendship.

During those days, Emerson developed an illness and Ann was worried about her husband's health. Traditional medicine hadn't helped and his condition was deteriorating, so Ann encouraged him to seek McMullen's help. Emerson wasn't sure about this, but he didn't have much to lose, so he sought his psychic friend's advice. Though George McMullen had only an eighth-grade education, he diagnosed the illness correctly and his advice proved to be helpful. It wasn't long before Emerson fully recovered.

Intrigued by his psychic ability, Emerson wanted to see if McMullen could help him with his archeological field work. So he took McMullen to previously known archeological sites and asked him to give as much detail as he could using his psychic abilities. McMullen, of course, had never been to any of these sites, nor did he know anything about them. Yet the descriptions he gave were so stunningly accurate that Emerson decided to explore unknown archeological sites with McMullen's help.

This psychic field work was conducted for two years before Emerson's 1973 address to the Canadian Archeological Association. One of them was a visit to the Iroquois prehistoric village of Quackenbush, about eighty miles from Toronto. Emerson described how information about this site was obtained. "George just sort of takes in the lay of the land, rapidly walking around, noting what is there today. Then he seems to become abstracted and begins talking about what he is seeing … as if it was as a functioning Indian village."[32]

By describing it as a "functioning village," Emerson meant McMullen could see the day-to-day activities of the villagers, and knew such details as the crops they consumed. He was certain the villagers' diet included corn, beans, and squash. According to the location of the village and based on pollen studies of the time period, Emerson's academic knowledge maintained that corn, beans, and squash were not cultivated there. However, upon further investigation it was found that these crops were available because of trade practices between the Iroquois and a distant tribe in a region where these crops were grown.[33]

Thus the unique partnership of Emerson and McMullen led to new archeological information and the establishment of psychic archeology in Canada. Schwartz included several other interesting accounts of psychic archeology in *The Secret Vaults of Time* that indicate the human mind in resonance with universal mind has access to what seems to be unlimited knowledge and information.

Operation Deep Quest

The concept of a universal nonlocal mind is so different from how we normally understand the nature of reality that when nonlocal phenomena like remote viewing are observed, investigators take precautions to make sure they have not missed explanations that would fit within the framework of classical physics. Stephan Schwartz had worked at SRI with Targ, and knew that remote viewing was nonlocal, which meant that distance, time, and energy were not involved in such observations. But he wanted to rule out "transmission" of remote viewing by means of extremely low-frequency (ELF) electromagnetic waves, which could have been a possibility consistent with classical physics. ELF waves are very long, on the order of miles, and can go through almost anything, unlike the short, high-frequency waves of radio and television transmissions. It was known that ELF waves couldn't function deep under the sea, so Schwartz, along with a team of psychics, scientists, and operators of the submersible *Taurus*, set out to do such an experiment. It was called Deep Quest. Psychics Ingo Swann, and Hella Hammid were asked to independently identify the locations and descriptions of sunken ships on an oceanic map covering fifteen hundred square miles.[34]

According to Schwartz, Swann and Hammid identified several wrecks on the map, all of which were known except one. Both remote viewers described this unknown site independently, and in the same way. Their descriptions included a sailing ship catching fire when its steam engine exploded and then sinking, approximately ninety years earlier. They also said there would be the aft helm of the ship lying with the wheel down and the shaft coming out of it, along with a steam winch nearby. In addition Hammid drew a sketch of a block of stone near the site measuring about 5 by 7 by 9 feet.

In the summer of 1977, Schwartz and his team dived into the waters near Santa Catalina Island off the coast of Southern California. Their mission was to find the sunken sailing ship based on the psychic information Swann and Hammid had provided. One of the team members of this expedition was a highly regarded scientist named Anne

Kahle. Her job was to control all of the records for this experiment to make sure that later, the information gathered could not be written off as inaccurate or fraudulent by the skeptics. In addition, they filmed the entire expedition.[35]

Initially as the search began in the murky waters, no shipwreck was found. Time was running out for Schwartz and his team, but the search continued, though hopes of finding the sunken ship diminished. To improve their chances of locating the wreck, a radio homing device was lowered in the location that Swann and Hammid had identified. As the captain turned *Taurus* in the direction of the homing device, its beeping sound was picked up. When they got closer, Swann pointed to an object on the ocean floor, sensing that it was part of the wreck. The robotic arm of *Taurus* was extended to pick up the object—and it was indeed a piece of the ship's wreckage. Soon, everything Swann and Hammid had described about the wreck was found, including the block of stone Hammid had sketched.[36]

A movie about this remarkable adventure, called *Psychic Sea Hunt,* is posted on Schwartz's website (www. stephanaschwartz.com). Before I watched this film I had read about this expedition in Schwartz's *Opening to the Infinite.* I hadn't expected to react emotionally as I watched *Taurus's* robotic arm lift a piece of the ship's wreckage out of the water, but I did, and I thought, *Pictures do speak a thousand words.*

Could this have just happened by chance? Not according to Don Walsh, who at the time was dean of the Coastal and

Marine Sciences Institute of the University of California, Davis. "We know submersibles," he maintained. "We know deep ocean engineering." Referring to Schwartz and his team, he added, "They would have had to beat us across the board. I'm just saying that this didn't happen by chance."[37]

While the experiment had been successful, and information about it had been observed, recorded, and maintained by an independent and well-regarded scientist, skeptics considered the results to be fabricated and the whole expedition a hoax.[38] However, as we will see with examples that follow, the fact that many mainstream scientists refute the capacity of human beings to see remotely doesn't invalidate its reality.

It had been determined at SRI that distance and time do not affect remote viewing, and once again, Swann and Hammid had shown the same. Working independently, they had arrived at similar information while thousands of miles away from the site of the sunken ship. How was it possible for them to know the exact location and details of a ship that sank approximately ninety years ago? Is it possible that all information and knowledge in the universe about the past, present, and future exist in the nonlocal universal mind, which psychics like Swann and Hammid can access? Is it possible for anyone to access the universal mind? And, if so, is there evidence for this? We will address these questions in the chapters that follow. For me, the need to find answers to these questions unexpectedly emerged decades ago in the lounge in North Dakota.

Universal Mind and Music

The phenomenon of connecting with universal mind can be deduced from the abilities of remote viewers. But there are other ways in which humans and universal mind seem to interact. In fact, as we will see, individual mind and universal mind exist as one in nonlocality, and their apparent separation is only an appearance in the local domain.

Let's review the story of a young American composer, Jay Greenberg, who began writing music at age two and had composed five symphonies by the age of twelve. He was born on December 13, 1991, in New Haven, Connecticut. His father, Robert Greenberg, who lost his eyesight at age thirty-six to retinitis pigmentosa, is a professor of Slavic languages at Yale University. His Israeli-born mother is a painter, and neither parent has a musical background.[39] Jay Greenberg's remarkable story was first aired on *60 Minutes* on November 28, 2004, when he was twelve. The following is an abbreviated description of the segment, hosted by Scott Pelley of CBS. It shows how Jay Greenberg composes music, which his brain seems to receive from the universal mind.

Greenberg likes to be called "Bluejay" because blue jays are small and they make a lot of noise. He is considered by some to be one of the greatest musicians of the last two hundred years.

At age ten he was awarded a full scholarship at Juilliard to study music. Here is how Jay's teacher, Samuel Zyman, who has taught there for seventeen years, describes Jay's musical ability: "We are talking about a prodigy of the level of the greatest in history, when it comes to composition. I'm talking about the likes of Mozart, Mendelssohn, and Saint-Saëns. This is not subjective, it's an absolutely objective fact—Jay could be sitting right here and he could be composing right now and finish a piano concerto, before our very eyes, in probably twenty-five minutes. And, it would be a great piece."[40]

Jay doesn't know where the music in his head comes from, but he is certain "It's coming by itself—It doesn't need to [be changed]." He adds, "The unconscious mind is giving orders at the speed of light and so I hear it as a smooth performance of a work already written." According to Zyman, Jay's ability to compose is unique. "It is as if he is looking at the picture of a score, and taking it from the picture." Jay composes music he sees in his mind's eye on his computer so fast that he often crashes the program.

Jay's parents were astounded not only by his musical ability, but by how he seemed to know musical instruments he had never seen, and how he could write music as a toddler without any music lessons. His mother says, "He just started writing at the age of two. He drew a cello, wrote the word cello, and asked for a cello. We were surprised, because neither of us has anything to do with string instruments. And, I didn't expect him to know what it was." Mr. and Mrs. Greenberg didn't own a cello, nor had Jay seen one before. Equally amazing was Jay's visit to a music store, where he saw a miniature cello for the first time. "He looked at the cello, sat down and started playing it," his mother says." She was amazed: "How do you know how to do this?" she asked him.[41]

Jay says it is not possible for him to shut out the music he hears, and that he receives several musical compositions at the same time. He explains, "There are multiple channels—it's what's been termed as my brain controlling two or three different pieces of music at a time. Along with the channels of everyday life and everything else."

Elizabeth Wolff, a well-known concert pianist who helped Jay with his piano technique, reported, "When Jay is bored he can play any piece of music backwards with the sheet music placed upside down." Jay demonstrates this ability with ease, to the amazement of Pelley and Wolff.

Jay's symphony *The Storm* was first played by the New Haven Symphony Orchestra when Bluejay was twelve year old. As Scott Pelley reports, "And when the last note sailed into the night, Jay navigated an unfamiliar stage to take a bow."[42] Almost everyone who saw this *60 Minutes* segment would be astonished by Jay Greenberg's musical ability, and many would consider him a child prodigy. However, it's also possible to see Bluejay's remarkable ability as a representation of the majesty, wonder, and oneness of human and universal mind.

Evidence of the Universal Mind in Identical Twins

On February 20, 1979, Thomas J. Bouchard at the University of Minnesota read a newspaper article about identical twins reared apart, which his colleague in the psychology department had left for him in his mailbox. A note attached to the newspaper said, "I think you will find this article interesting." As soon as Bouchard finished reading it, he wanted to conduct a study on identical twins. The resulting study, titled "Minnesota Study of Twins Reared Apart (MISTRA)," began in March of 1979 and soon gained worldwide attention, not only among academics and clinicians but also among the public at large.[43]

I remember reading this article about the "Jim twins" in the *Minneapolis Star Tribune* and being astonished by their similarities. Some of my friends who had read the report felt the same way, and it triggered lively discussions between us about the role of "nature and nurture" in human personality and behavior. Even back then I couldn't see how just nature or nurture could possibly explain some of the remarkable coincidences in the lives of the Jim twins. However, at the time, the routines and responsibilities of life pushed aside my curiosity. Many years later as I began work on this book, I remembered the unique and uncanny similarities between the twins, so I purchased a copy of *Born Together—Reared Apart: The Landmark Minnesota Twin Study* by Nancy L. Segal to see if I could find better answers to the seemingly inexplicable similarities of identical twins than those my friends and I had discussed years ago.

As Segal reported, their striking similarities included first names (Jim), their favorite school subject (math), their dreaded school subject (spelling), their preferred vacation spot (Pass-a-Grille Beach in Florida), their past occupation (law enforcement), and their hobbies (carpentry). They both named their sons with the first name James and variations of the same middle name (Alan and Allan). At age thirty-nine, both of them gained ten pounds at the same time for no apparent reason. They had similar headaches, which first started for them at age eleven. Their weights and heights were almost identical. Jim Lewis weighed 154.90 pounds and was 70.90 inches tall, while Jim Springer weighed 154.59 pounds and was 71.40 inches tall. Their vocational test scores were similar, and their (nonverbal) IQ scores were one point apart.[44]

Amazingly both were married twice, the first time to wives named Linda, and the second time to wives named Betty. Both of them owned a dog named Toy. They chain-smoked Salem cigarettes, drove Chevrolets, loved stock-car racing, chewed their fingernails, and liked spicy food. Another uncanny similarity was that they both flushed the toilet before using it. They both spoke of feeling emptiness in their lives, which disappeared after they met one another. They both described their relationship as "closer than best friends."[45]

Bouchard and his colleagues leaned toward a genetic explanation for the remarkable similarities of these identical twins reared apart. But there were too many similarities between the Jim twins to possibly be solely genetic in nature. In fact, explaining human nature strictly in the framework of genes or the environment while ignoring the larger influence of a universal mind seems unreasonable. In an interview with the *New York Times*, Bouchard, the lead researcher for the study, said, "If someone else brought the material to me and said, 'This is what I have got,' I'd say I didn't believe it."[46]

Consider how the Jim twins named their sons: Are there specific genes for choosing names? Did genes have anything to do with marrying women named Linda in their first marriages, and wives named Betty in their second? Why did both have dogs named Toy? How could nature or nurture explain their common interests in stock-car racing and driving Chevrolets?

It's difficult to miss the evidence of mind-to-mind connection and the existence of a universal mind in these uncanny similarities. In fact, greater insight from such similarities also suggests that though the universal mind maintains its mystery, it also reminds us of its presence. Why? Perhaps so we may find ways to connect to it for purposes of greater knowing. The evidence for the universal mind and our connection to it has profound implications. It tells us about the role and influence of the universal mind on our thoughts, words, and deeds. It also tells us we are not simply influenced by the universal mind, but that we influence it as well. It is a two-way feedback loop. Our thoughts and feelings in conjunction with our destiny co-create the realities we experience.

Destiny includes examples like these: All of us can learn how to play music reasonably well, but not everyone can compose symphonies at the age of twelve like Jay Greenberg can. We can all become better at math by studying and working hard, but most of us will not be able to multiply large numbers or give their square root as well as the savants who can do so effortlessly, as we will see in the next section. The ability to access the universal mind is predetermined, but we can also learn how to access it consciously. The practices of meditation, visualization, intent, desire, and being in alignment with the design of the universe allow us such access. In many cases, we can learn to access the universal mind without significant effort, as has been demonstrated by Russell Targ in his teaching of remote viewing. Another example of our ability to access universal mind is by paying attention to the coincidences in our life. Greater awareness of the coincidences we experience accentuates the possibility for more coincidences to occur.

The fascinating story of the Jim twins brought many others to MISTRA, a study that lasted for twenty years. At the end of the research project, 137 reunited twins had been studied, resulting in more than 150 publications. Among many other twins who came to the University of Minnesota, two middle-aged British women named Bridget Harrison and Dorothy Love were particularly interesting. These twins were separated in infancy during World War II and raised apart in different socioeconomic settings. When they got off the plane in Minneapolis, they were each wearing seven rings, two bracelets on one wrist, and a watch and a bracelet on the other wrist. Bridget had named her son Richard Andrew, and Dorothy had named her son Andrew Richard. Dorothy's daughter was named Karen Louise, and Bridget had named hers Catherine Louise.[47]

Even an ardent materialist may be reluctant to suggest that there are specific genes for wearing identical jewelry, in the same manner and on the same day. Nor could materialism explain why twins reared apart would give their children similar names even though they had different backgrounds—though materialists are not shy about using the dumping ground of "chance" to explain the remarkable similarities between Harrison's and Love's thoughts and actions. However, others find nonlocal, mind-to-mind connection or the reality of universal mind to be more compelling explanations.

Savants and the Universal Mind

Savant syndrome is a rare condition in which persons with serious mental handicaps resulting from mental retardation, autism, or major mental illness simultaneously possess brilliance within a narrow range of human skills. Most commonly these include prodigious memory; exceptional abilities in music, art, and sculpture; mechanical or spatial skills, including the capacity to measure distances precisely without the benefit of instruments and mastery of maps and direction finding; and mathematical skills including calendar calculating and computing large numbers at lightning speed. Other skills reported less often include exceptional language (polyglot) facility; unusual sensory discrimination in smell, touch, or vision, including synesthesia (i.e., seeing alphabets represented in colors, or the feel of different objects having particular smells associated with them), as well as extrasensory perception. Usually only one skill exists, but in some savants several skills exist simultaneously.[48]

How can savants answer complex and elaborate mathematical questions without knowing what a formula is? How can Jeremy, who is severely autistic and cannot count, stand beside the railroad tracks and give you the total number of boxcars as the train rolls by? How can Leslie Lemke, who was blind, palsied, and mentally handicapped, play Tchaikovsky's first piano concerto flawlessly at age fourteen after having heard it just once? How can Alonzo Clemons, with poorly developed speech and an IQ of 40, sculpt the animals he sees from memory, after only a passing glance? Darold A. Treffert, MD, an expert of the savant syndrome, in his book *Extraordinary People: Understanding Savant Syndrome*, has written extensively about people like Jeremy, Leslie, and Alonzo. He suggests that his findings have far-reaching implications regarding buried potential in some—or perhaps all—of us.[49]

Like many people around the world, I first came to know about savants when I watched the award-winning 1988 film *Rain Man*, which Treffert consulted on. Dustin Hoffman played the role of Raymond Babbitt, an autistic savant, whose fictional character was inspired by the real savant Kim Peck (1951–2009). Peck was a mentally handicapped man who had an encyclopedic knowledge of music, literature, history, sports, and nine other areas of expertise.

Kim's father referred to his son as a walking-talking "Google." It was an appropriate description. He memorized more than six thousand books and could name all the US area codes and major-city zip codes. He memorized the maps at the fronts of telephone books and could tell you precisely how to get from one US city to another, and then how to get around in a city street by street. His calendar-calculating skills regarding past and future dates, and the days of the week they fell on, were lightning fast. Later in life he developed musical abilities as well. He could read a page with one eye and the adjacent one with the other eye and have perfect recall of what he read.[50]

Another savant, one Treffert claims was the most remarkable he ever met, was Leslie Lemke. He was born in Milwaukee, Wisconsin, on January 31, 1952, and his mother gave him up at birth. In spite of his triple handicaps of blindness, retardation (a term used by Treffert without negative connotations), and cerebral palsy, he played Tchaikovsky's first piano concerto flawlessly at age fourteen, having heard it for the first time on television. From there, Leslie moved on to improvisation, having become bored with just reproducing what he heard. Instead, he created new songs, which he composed, played on the piano, and sang.[51] Recently, watching a video of Leslie playing and singing "How Great Thou Art" brought tears of wonder and joy to my eyes.

Treffert's suggestion that the remarkable abilities of savants may be hidden capacities among all human beings is not merely speculative. Actress and five-time Golden Globe nominee Marilu Henner remembers every single day of her life in exquisite detail, including the clothes she wore, who she was with, how she felt, what she ate, and what the weather was. She's not alone. There are at least a dozen or more people documented in medical literature who live normal lives and can remember every person they have ever met, every conversation, and every possible detail about any day of their lives. These people are referred to as having highly superior autobiographical memory (HSAM). Lesley Stahl interviewed Henner and five others with her ability on *60 Minutes* on December 19, 2010. When asked if the tremendous quantity of information in their brains interferes in any way in their thinking, they all said no.

When asked how they could recall vivid details about every day of their lives, they answered, "We see it is as a video clip in our mind."[52]

Deepak Chopra, MD, and Rudolph Tanzi, PhD, in their book *Super Brain: Unleashing the Explosive Power of Your Mind to Maximize Health, Happiness, and Spiritual Well-Being*, argue that we have set low norms for our mental powers. They point to hyperthymesia (remarkable memory), hyperlexia (exceptional reading ability), and hyperpolyglots (extreme ability to learn foreign languages) as examples of the supreme capabilities of our minds. They suggest, "There is no reason to regard exceptional performance as excessive, a word that implies something freakish if not disordered," and add, "Our view is that we could be evolving into a new norm higher than ever before." They provide numerous examples of how our choices create new neural pathways and synapses, along with new brain cells, causing us to undergo a second kind of evolution that rests upon personal choice.[53]

Among many examples of our minds being limitless, Chopra and Tanzi point to Timothy Doner, who decided to learn Hebrew in 2009, soon after his bar mitzvah. With the help of a tutor, he soon became good at it. Discussing Israeli politics with his tutor led him to think about learning Arabic, which is regarded as one of the five hardest languages to learn. When he attended a university course to study it, it took him a mere four days to learn the alphabet and a week to read it fluidly. He then went on to learn Russian, Italian, Persian, Swahili, Indonesian, Hindi, Ojibwa,

Pashto, Turkish, Hausa, Kurdish, Yiddish, Dutch, Croatian, and German, all of which he taught himself from grammar books and flashcard applications on his iPhone.[54] You can find him speaking in many different languages with ease in videos online (https:en.wikipedia.org/wiki/Timothy_Doner).

I know Hindi, so I wanted to see how well Doner spoke it. Especially because he had learned it in a matter of days, I was impressed not only by the grammar he used, but his accent.

As interesting and unusual as people with extraordinary abilities are, in the context of a universal mind, they are not. But if we are connected to the universal mind, why doesn't everyone know everything? At the conscious level we don't have direct access to the universal mind, but at the unconscious level we not only have access to it; we *are* the universal mind. All of us have the *power within*. The notion that each of us is "super human" seems to defy logic only because we have not properly accounted for or understood the existence of a universal mind.

We have assumed that memory, intellect, and ESP originate and reside in our brains. According to biologist Rupert Sheldrake, metaphorically, our brains are like a television set that receives a signal from a broadcasting station and converts it into pictures we see on the screen. When the TV is broken, it is unable to receive the broadcast signal, but that doesn't mean the signal is not being sent. In the same way, the knowledge and information we experience in our brains and our bodies are received by the "transmissions" of the universal mind. Unlike the television set, which only receives signals, human beings have free will, egos, intentions, desires, hopes, dreams, and aspirations, and these attributes influence the universal mind. It is a two-way signal. This is why meditation, prayers, intentions, and beliefs often affect the outcomes and realities we experience. Further discussion of meditation, prayer, and our ability to interact with consciousness will be discussed in chapter 6. Greater accessibility to the universal mind lies dormant for most of us, yet unexpectedly at times, when we need it most, it becomes available in the form of ESP, as it did for me on my first night in North Dakota.

The existence of a universal mind is also at the heart of spontaneous healing and the placebo effect. Prodigious musical and mathematical skills and encyclopedic memories don't present themselves to most of us under normal circumstances, but better access to the universal mind in some of us suggests that it may be possible for everyone to have greater access to it. Seasoned meditators or people under hypnosis, for example, may gain information that is far beyond the capacity of the conscious mind.

Ancient wisdom coupled with the concept of entanglement in quantum physics strongly suggest that the universal mind is real. It is the source of all phenomena in the known universe. Even though universal mind is nonlocal, it is an integral part of the local domain. In other words, the universal mind is who we are. In ordinary everyday life, we experience universal mind in everything we see, hear, know, or do. From the beating of our hearts to the creativity of our minds; from every breath we take to every word we speak;

from the love we experience to the things we discover: they all take place because of our two-way interactions with the universal mind. In the next chapter we will discuss how our universal mind creates matter.

4

Mind Matters—How Thoughts Create Reality

A human being is a part of a whole, called by us "universe," a part limited in time and space. He experiences himself, his thoughts and feelings, as something separated from the rest—a kind of optical delusion of his consciousness. This delusion is a kind of prison for us.... Our task must be to free ourselves from this prison.
—Albert Einstein

E instein's message is sound advice. But how can we free ourselves from the prison of delusion? When we realize that our notions of reality may be flawed, and learn to broaden our perspective, we may realize who we really are—and recognize our supreme nature.

This awareness permits us to live in accordance with the truths evident in new sciences and ancient wisdom. In this chapter we will discuss the theoretical, experimental, and observational evidence for how thoughts create reality. Such an understanding will help us free ourselves from our "optical delusion." We will discover, as astrophysicist Sir James Jeans informs, "The stream of knowledge is heading towards a non-mechanical reality; the universe begins to look more like a great thought than like a great machine. Mind no longer appears to be an accidental intruder into the realm of matter ... we ought rather hail it as the creator and governor of the realm of matter."[1]

Some of us are familiar with the concepts of visualization, affirmation, goal setting, and meditation, and how these techniques produce tangible outcomes by using the power of our thoughts. Furthermore, numerous self-help books explain how we can use these techniques to enhance or exceed our hopes, dreams, and aspirations. Perhaps most commonly, we have seen athletes use these methods to improve their skills and performance. The relationship between thought and matter has emerged into our awareness only recently, but the idea has been around for a long time.

In Greek mythology, Pygmalion was an exceptional sculptor from Cyprus who sculpted the most beautiful woman he had ever imagined out of ivory. So exquisite and lifelike was the statue that he fell in love with it, and he wished for a bride in its likeness. He kissed the statue, the woman came to life, and Pygmalion's wish was granted— they married.

In ancient Greece mind-matter interactions were considered so real that there are several other mythological depictions of statues coming to life.[2] But depictions of the interconnectedness of mind and matter are not limited to Greek mythology. Ancient Indian philosophy explains that all matter and observable phenomena are reflections of consciousness: in other words, matter emanates from the universal mind.

Today, we are charmed by Pygmalion's story but generally consider it to be metaphorical. In this chapter, we will review accounts of mind-matter interactions to show that mind *creates* matter. In reviewing the placebo effect, the body-mind, and evidence of mind-matter interactions, we will discover that the myth of Pygmalion was not just metaphorical. It was a representation of a deeper truth that modern science has shown to be literally true.

Let's begin with what is known as the "Pygmalion effect" in the social sciences, and how it came to be. Harvard psychologist Robert Rosenthal, in his earlier academic career (1957–62), taught in the psychology department at the University of North Dakota. There, he conducted a clever experiment to test the power of human *expectation* on lab rats.

Rosenthal instructed a group of lab researchers to experiment with rats that he said were specially bred to be either exceptionally good or exceptionally bad at running mazes. The truth was that none of the rats had been bred in any way; they were simply ordinary lab rats. Rosenthal posited that human thought alone would affect the behavior of rats, and sure enough, as he predicted, the "bright" rats ran mazes faster and more accurately than the "dull" rats did. The experiment demonstrated that the subconscious expectations of the researchers had indeed influenced the performances of the rats. Rosenthal published this finding in a 1963 article in *American Scientist* in which he speculated that "if rats become brighter when expected to, then it should not be farfetched to think that children could become brighter when expected to by their teachers."[3]

Rosenthal found an opportunity to test his idea at Spruce Elementary School in South San Francisco. In the double-blind study he created, neither the children, nor their parents, nor their teachers knew about the experiment he was conducting. The only information teachers were given was that a small group of children in each classroom was poised for intellectual growth. At the end of the school year, students whose teachers *expected* them to bloom intellectually increased their IQ scores by more than 27 points. These students, remember, were no different from any of the others, simply chosen to test Rosenthal's hypothesis. And indeed, within a year the subconscious, nonverbal expectations of the teachers made the students smarter.[4]

Rosenthal published these findings in his book *Pygmalion in the Classroom: Teacher Expectation and Pupils' Intellectual Development*, coauthored with Lenore Jacobson in 1968. (For reasons of privacy, Spruce Elementary was referred to as "Oak School" in the book.) The implications of this study were far reaching and represented one of the most inspiring breakthroughs in the history of psychology, but for a long time Rosenthal's findings were lambasted by prestigious academicians and researchers who asserted that the study was flawed and the conclusions were ridiculous.[5] In 1978, Rosenthal and his colleague Donald Rubin summarized 345 experiments involving the influence of interpersonal expectations, and concluded, "The reality of the phenomenon is beyond doubt." Rosenthal's key assertion regarding the Pygmalion effect was that expectations had to be *subconscious*, as the conscious expectations that had been studied by others had not produced similar outcomes. Eventually, criticism faded and the Pygmalion effect became part of mainstream thinking throughout the world. In fact, it is also referred to as the "Rosenthal effect."[6]

Though the Pygmalion effect has been accepted in mainstream psychology and the social sciences, it is normally explained in the limited context of the psychosocial framework. Better intellectual achievements of children resulting from higher teacher expectations are attributed to their positive interactions, such as the way teachers speak to them: their encouraging facial expressions and friendly body language. While these psychosocial explanations are valid, they do not represent Rosenthal's central finding—that children grow smarter *because teachers think they are*. The significance of how a teacher's thoughts improve a child's brain (matter) may not be readily accepted in mainstream thinking, but there is ample evidence that their thoughts do.

For example, let's review how a teacher's high expectations of her grade school students in a poor neighborhood in Chicago led to their performing exceptionally better than other children of similar backgrounds.

Marva Collins's Story

In the late 1960s while teaching at a public school in Chicago, Marva Collins believed that low teacher expectations for poor black and other minority children were at the heart of their subpar academic performance. She tried hard to convince teachers and administrators there of the critical role of teacher expectations on student performance, but to no avail. So, unwilling to accept low standards and expectations for such children, and with the support of her husband, Collins opened a one-room school in her own house, which she named Westside Preparatory School.

Collins had often seen the children enrolled in the public school where she taught labeled as having learning disabilities, put in special education classes, and assumed to have poor IQs based on culturally biased exams. Yet when some of those children enrolled at Westside, they performed exceptionally well. What was Collins's secret for their success? Were her high expectations of their intelligence the reason? Collins believed that children have innate capacities to learn that are greater than what is generally assumed, and that when adults encourage and expect them to succeed, they do. And she was right; her students' academic achievements were far beyond the ordinary, and in the 1980s, she was regarded as one of the best teachers in America. Yet despite this recognition, critics asserted that the academic achievements Collins's students had attained wouldn't necessarily equate with success in life as adults.

On the first day of school, Collins would begin by telling

students that she expected them to learn and succeed. "I know most of you can't spell your name. You don't know the alphabet, you don't know how to read, you don't know homonyms or how to syllabicate.

"I promise you that you will. None of you has ever failed. School may have failed you. Well, goodbye to failure, children. Welcome to success. You will read hard books in here and understand what you read. You will write every day so that writing becomes second nature to you. You will memorize a poem every week so that you can train your mind to remember things."[7]

Though on their first day of class the children did not know how to read or write, Collins began by handing out copies of "Self-Reliance" by Ralph Waldo Emerson. And when the kids looked confused, she told them, "I don't expect you know how to read this. I will read it to you."[8] Then, not only did she read the essay, but she engaged the students' curiosity by telling them in detail who Emerson was, where he was from, and what he was like, and about the work he did. After telling them he was from Boston, Massachusetts, she gave them a brief geography lesson about that city and state.

In her efforts to encourage the children to study hard, she told them that Emerson had studied hard as a youngster, diligence that had permitted him to attend Harvard College. She quoted from "Self- Reliance," explained what Emerson meant, and asked questions that she answered for the students. Then, to encourage class participation, she asked the same questions again so the students could repeat the answers she had provided.

This was on their first day of school!

Later, Collins would teach from Plato's *The Republic*, Homer's *The Odyssey*, and other works including *Little Women, Candide, Charlotte's Web*, and *The Brothers Karamazov*; she meant it when she told the children they would be studying "hard books." She also maintained high standards and expectations while teaching math, science, and social studies. Her style was informal, engaging, constructive, and full of encouraging comments for each of her students, and year after year, they achieved high academic success. As word got around, the press started covering her story in articles in newspapers and magazines. In 1979 a *60 Minutes* interview with Morley Safer featured Collins and her students. I happened to catch it, and it was evident to me that she believed in her students' success.

Their performance and her philosophy of self-reliance caught President Ronald Reagan's attention in 1980, and she was invited to serve as Secretary of Education in his administration. Earlier, she had been asked to serve in the same capacity by President Jimmy Carter as well, but she turned both of these requests down in favor of continuing teaching at Westside. In 1982 she published a book about her work titled *Marva Collins' Way*, which brought her method to the attention of the public at large. Teachers and administrators from all over the United States and around the world came to attend her workshops and seminars.

In 1981 her story was made into a movie in which Cicely Tyson played the role of Marva Collins, and Morgan Freeman her supportive husband. Thus in 1995, sixteen years after the first broadcast of Collins's interview on

60 Minutes, her story was presented again. This time her grade school students were young adults who held good jobs in various careers, proving wrong the early critics who maintained that Collins's success with children would not last into adulthood.

Collins's encouraging, affirming, and loving approach to interacting with her students made them believe they could learn. She connected with them, consciously and subconsciously, through constant reminders that they were special, smart, and capable. High expectations were at the heart of the students' academic achievements, and were accepted as such by teachers and administrators around the country. But in practice, Collins's method was never fully implemented in other schools in the US.

Though teachers' expectations had been accepted as a significant reason for high student achievements, the *literal effects* of teachers' thoughts and feelings on students' brains were not. Today, though, we know about neuroplasticity and how the brain's neuro networks constantly change and react to every thought and feeling. We also know much more about the placebo effect, psychoneuroimmunology, and body-mind phenomena. These scientific breakthroughs have made it possible for us to understand that Collins's expectations had *improved her students' brains.*

The Power of Thought

There is considerable evidence in science that our thoughts create reality! Consider the evidence we now have from the placebo effect, the mind-body connection, the manner in which our thoughts influence our genes, and the behavior of subatomic particles. All these phenomena lead to one conclusion: we can use the power of thought and belief to construct the reality we wish—when these expectations are held subconsciously. The reality we wish may take various forms, including good health, achievements in our work, meaningful relationships, overall well-being and happiness, or even transcending linear time. Time in the local domain or in the physicality of the 3-D (three dimensional) world is experienced as past, present, and future. Yet even in the physical domain, it is possible to experience time in the nonlocal realm of *no-time*, or simply the present. Amazingly, we can do this through the power of our subconscious thoughts and beliefs. It may seem remarkable that we could influence linear time; the condition of our health and well-being; or the realities we wish to experience through what we think and believe. But let's see how Ellen Langer illustrates all this to be true.

Ellen Langer's *Counterclockwise*

Ellen Langer, PhD, is a modern-day Pygmalion. As a professor of psychology at Harvard University in 1979, she studied the effects on a group of eight elderly men of turning back the clock *psychologically* to determine whether they would experience any corresponding *physiological* effects. In 2009 she published *Counterclockwise: Mindful Health and the Power of Possibility*, in which she described the details of her study. I have summarized her research here, which shows how the thoughts and beliefs of men in their seventies and eighties reversed their aging—in just one week![9]

Langer interviewed several men and chose sixteen to participate: eight in the experimental group and eight in a control group. The men in the experimental group were told to imagine their lives in 1959 as if they were experiencing that year in the present. To help make that year feel real to them, she found an old monastery in Peterborough, New Hampshire, and retrofitted it in the decor of the 1950s, including choices of furniture, carpeting, curtains, and decorative artifacts. Furthermore, to help them immerse psychologically in a time when they were twenty years younger, she provided books, magazines, music, and movies from the '50s as well. The men were also instructed to converse with each other as if it were 1959. Soon after the experiment began, several of the men in the experimental group started cooking, cleaning, and doing some of the other physical activities they had done twenty years earlier.

A week later the control group of eight men was invited to spend a week in the same facility. They lived in the same rooms, ate similar food, and were encouraged to interact with each other the same way as the experimental group had. But these men were asked only to *reminisce* about the past. After the weeklong retreat, the men in both the experimental and control groups were tested for physiological changes.

Langer and her graduate students found improvements in the memories and hearing capabilities of both groups. However, the experimental group showed a greater improvement in joint flexibility, finger length (arthritis had diminished), and manual dexterity. On an intelligence test,

63 percent of the experimental group improved their scores, while 44 percent in the control group did. Improvements in height, weight, gait, and posture were also observed. When observers who didn't know the purpose of the study viewed before-and-after pictures of the participants at the end of the study, all of the men in the experimental group were judged to be noticeably younger than those in the control group.

This experiment illustrates not only the mind-body connection, but also that thoughts combined with feelings and beliefs can transcend linear time. Thoughts and beliefs can transform our physical reality to the time we imagine ourselves in. When our hearts and minds bring the past or future into the present, we experience those different times.

Another 2007 study conducted by Langer with Alia Crum of Harvard University found that hotel maids who got the same level of exercise received different benefits from it depending on whether they believed their work was exercise or not. The study involved eighty-four hotel maids whose jobs actually provided enough exercise to exceed the US surgeon general's recommendation for daily exercise. These women cleaned, on average fifteen rooms a day, each taking between twenty and thirty minutes to complete and requiring lifting, bending, reaching, and pushing. But the women didn't know this meant they were getting exercise, and most didn't see themselves as physically active. In fact, when they were asked about the amount of their daily exercise, one-third didn't believe that they got any at all and two-thirds reported not exercising regularly.

For the experiment, the researchers split the maids into

two groups and recorded their physiological measurements, such as weight, body mass index, and blood pressure. The women in the experimental group were told about the number of calories they burned while lugging equipment around, vacuuming, changing bedsheets, and general cleaning. They were also told their work exceeded the surgeon general's recommendations for daily exercise. The women in the control group performed all the same activities but were not given this information. After a month, Langer and Crum found that the experimental group had lost weight and decreased their waist-to-hip ratio, body mass index, and body-fat percentage, and that their blood pressure dropped by 10 percent. No changes were observed for the women in the control group. Women who thought they were exercising experienced physiological changes, while others who did the same work but didn't consider it exercise had not. Langer's experiment suggests it is possible to trigger the placebo effect with words.

The Powerful Placebo

The placebo effect can be seen as a "first cousin" to the Pygmalion effect because both phenomena demonstrate how our minds create material reality. *Expectation* and *belief* are essential in achieving any goal or outcome, and they are at the heart of the mind-body connection. When subjects expect and believe that a blue pill (placebo) will cause hives, it does. When this same blue pill is given with the instruction that it will heal hives, it does that too. The blue pill itself is inert, and therefore does not cause any reaction. Expectation and belief in what the pill can do causes the outcomes.

Though exactly how the placebo works is not known, research on the topic over the last couple of decades has been extensive, and today we have gained a broader understanding about it. Fabrizio Benedetti, a world authority on placebos and a neuroscience professor at the University of Turin Medical School, states, "Placebos are made of many things, such as words, rituals, symbols, and meanings. Thus, a placebo is not the inert treatment alone, but rather its administration within a set of sensory and social stimuli that tell the patient that a beneficial therapy is being given. Indeed, a placebo is the whole ritual of the therapeutic act."[10]

Benedetti is not alone in asserting that words, rituals, symbols, and meanings influence the placebo effect. Ted Kaptchuk of the Harvard Medical School, another expert on placebos, explains, "When you look at studies that compare drugs with placebos, there is an entire environment and ritual factor at work. You have to go to a clinic at certain times and be examined by medical professionals in white coats. You receive all kinds of exotic pills and undergo strange procedures. All this can have a profound impact on how the body perceives symptoms because you feel you are getting attention and care."[11] This was true in Langer's experiment of age reversal in which she had used a set of sensory and social stimuli to trigger the powerful placebo effect, which is supportive of Benedetti and Kaptchuk's findings.

In a classic study, patients were given a placebo they believed to be a real drug for Parkinson's disease (levodopa).

Functional magnetic resonance imaging scans of these patients showed that a part of their brain lit up when patients ingested the placebo as it would with a real drug. Symptoms of this disease arise from impaired production of dopamine in part of the brain, which affects movement, and patients who were given a placebo but told it was an anti-Parkinson's drug were better able to move. Scans showed that the brain was activated in the area that controls movement, and the placebo boosted the patients' dopamine by 200 percent. The patients' improved maneuverability wasn't just a figment of their imagination. The power of thought and belief produced a physical change in their brains. This experiment has been replicated in numerous trials.[12]

Thousands of studies on the placebo effect show that almost any disease can be cured by the power of our thoughts and beliefs, but some experts like Kaptchuk think that while placebos help alleviate symptoms, they do not cure disease. He writes, "Placebos may make you feel better, but they will not cure you. They have been shown to be most effective for conditions like pain management, stress-related insomnia, and cancer treatment side effects like fatigue and nausea." Though Kaptchuk overlooks the vast body of evidence that shows healing of practically all diseases, nonetheless he encourages triggering one's own placebo effect for better health. He explains, "Engaging in the ritual of healthy living—eating right, exercising, yoga, quality social time, meditating—probably provides some of the key ingredients of a placebo effect."[13]

Kaptchuk's suggestions for triggering the placebo effect

and effective. However, there are many other ...ul ways by which our thoughts and beliefs create the realities we desire. Techniques like affirmation, visualization, chanting, hypnosis, and biofeedback have similar effects as a placebo. Let's continue to review the scientific evidence of mind-matter interactions.

Intellectually, we know that placebos work when we believe they do. However, normally in our daily life, we are so conditioned to believe that it isn't possible to cure an illness or physical ailment using our thoughts and beliefs that we reject the idea. How can we convince ourselves that it is possible to trigger the placebo effect without taking a pill or being primed with words or rituals, as experts suggest? Personally, reviewing the scientific evidence and attaining a greater understanding of how mind and matter interact have helped me lose weight, gain muscle mass, and overcome a few health issues.

One of the most significant and compelling uses of the mind to affect reality is illustrated by Dr. Joe Dispenza's experience.

Remarkable Recovery

In 2004, I saw the movie *What the Bleep Do We Know!?* In it, several scientists offer explanations for how our thoughts and feelings create the experiences of our lives. Actress Marlee Matlin plays a young woman named Amanda who discovers she can transform her life when she becomes aware of the quantum field hidden behind her everyday reality. One of the featured scientists in the movie is Joe

Dispenza, DC, who is also qualified in the fields of bio-chemistry, neurology, neurophysiology, and brain function. His comments in the movie about the powers of the human mind and how thoughts create reality captured my attention, and I wanted to learn more.

So three years later when he spoke at the Temple Israel in Minneapolis, I attended his presentation. Every seat in the large synagogue was occupied: there must have been fifteen hundred people or more, and the anticipation of the crowd was palpable. Dispenza had come to Minneapolis to promote his book *Evolve Your Brain: The Science of Changing Your Mind*, which had just been published.

He began by sharing that the inspiration for writing his book was a traumatic, life-threatening experience he had in 1986, when he was twenty-three years old. To capture the essence and accuracy of the story he told at Temple Israel, I have included information from his book, which documents his accident and recovery at length.

Dispenza had just started his chiropractic practice in La Jolla, California, and was living a carefree, high-octane life while training for a triathlon in Palm Springs. When he started the twenty-six-mile bike portion of the race, he was feeling good. He was familiar with the path, and he was clipping along when he was hit by a red SUV traveling at 55 miles per hour. He was thrown off his bike and, after flying several feet in the air, landed, bounced, and rolled uncontrollably for about twenty feet. As he lay on his back, he felt blood pooling in his rib cage, and he knew that something was seriously wrong.

He was taken to a hospital, where, after twelve hours

of X-rays, CT scans, and several other tests, the attending orthopedic surgeon told him the neurological examination was normal. But the X-rays showed that the mid-spine vertebrae, T8 through T12 and L1, were compressed, fractured, and deformed. When Dispenza saw that T8 was more than 60 percent collapsed, he thought it could have been worse: His entire spinal cord could have been severed. He could've been paralyzed or dead. Being a chiropractic doctor, he knew that the standard procedure for an injury like this would be a thoracic laminectomy with Harrington rod surgery.

Dispenza explained what this surgery meant: "The surgeon employs a tool box of carpenter's blades and mini circular saws to cut away the bone and leave a smooth working surface. Next the surgeon inserts the Harrington rods, which are orthopedic stainless steel devices. These are attached with screws and clamps on both sides of the spinal column to stabilize the severe spinal fractures."[14] The surgeon informed him that the rod would be eight to twelve inches long, and he was asked to pick a day for surgery within the next seventy-two hours. He was also told that the procedure would leave him paralyzed from the chest down. To emphasize the importance of this surgery, the surgeon added that he had never heard of a patient in the United States who had opted not to have one.

After consulting three other surgeons, who also recommended surgery, Dispenza agonized about his difficult choice. As a chiropractic doctor he thought that he, too, would recommend surgery if one of his patients had a similar medical condition, because it would be the safest option. But he didn't feel right about surgery for himself. Chiropractic philosophy had taught him that the innate intelligence that gives life to the body also heals it. He also realized that the four surgeons he had consulted understood life differently. So he checked out of the hospital, somehow knowing he was making the right choice.

With the help of friends, he moved into a beautiful, multi-window, spacious A-frame house and began his healing process. He used self-hypnosis, meditation, and affirmation for three hours every day. He visualized being healed and imagined his spine was normal. He mentally reconstructed his spine, rebuilding each segment of it. He viewed hundreds of pictures of spinal cords to perfect his mental imagery, and believed that a greater intelligence, which is in every one of us, was helping him heal.

Dispenza knew that the force of gravity stimulates calcium molecules to deposit on the surface of the bone, which helps heal it. So he asked his friend to build an incline board on which he could lie to speed his recovery. In six weeks he had recovered enough to begin swimming in the pool at the house. At eight weeks he could crawl, and could do yoga postures while lying down.

Dispenza's continued recovery led him to realize and fully accept that our thoughts and beliefs influence the nature of reality. He believed he was responsible for everything that had happened in his life, including his injury. Ten weeks after he began his intensive mind-over-matter healing, he returned to work and was able to engage fully in all aspects of his life.

Dispenza's miraculous recovery is important not only

because it is unique and inspirational, but also because it has far-reaching implications about who we are as human beings and what we are capable of. Thoughts, along with beliefs, focused attention, and clear expectations and intentions, link us to the infinite capacities of the universal mind, and can produce the outcomes we desire.

The reason we don't normally experience the kind of recovery that Dispenza did is our inaccurate understanding about the nature of reality. When we broaden our awareness and accept that the universe is "more like a thought," to paraphrase Sir James Jean, we realize that the individual and the universal mind are interconnected, and that we are infinitely more than what we assume ourselves to be. In the next chapter we will review some of the concepts normally understood to be true and show that new scientific evidence and ancient Eastern philosophy paint a different picture.

After his inspiring talk, Dispenza signed copies of his book. I stood in line to have him autograph mine. When I reached the table where he was seated, he looked at me and said, "I picked up on your energy ... I saw your smile and the way you were listening ... thank you!"

I was surprised. "You saw me in the crowd?" I asked.

He smiled. "How could I not?" He autographed my book with these words, "Nuri—your heart is open—that's what opens the mind." I was touched by such an unexpected encounter, but didn't realize back then that my intentions and thoughts about learning as much as I could about how thoughts create reality might in fact have helped create my brief but memorable moment with Joe Dispenza.

Though Dispenza's recovery was amazing, and I was

moved by it, I was still in the "awe and wonder" phase of how our thoughts cause material change. It took several more years of research, reflection, and contemplation, and a few psi experiences, for me to better understand who we are, and what is possible and why. This greater understanding convinced me that it is not only possible to live a healthier life, but also a happier, more meaningful, and more joyous one.

How did Dispenza recover? He made several conscious decisions and actions—such as doing visualization, affirmation, meditation, self-hypnosis, and yoga—but a key component of his recovery was the fact that *fields*, which are templates, shape and form all living and nonliving matter. Dispenza's recovery was a result of his belief that he could reconstruct his spine, based on what he knew and had accepted subconsciously. Without his knowing how it works, the perfect image of a spine, which exists in the form of a "blueprint" in the field, helped shape Dispenza's perfectly normal spine in actuality.

What Are Fields?

The evidence of fields is readily observed when a magnet is placed under a sheet of paper and iron filings are placed on top. The magnetic force underneath the sheet arranges the iron filings in a pattern. When the iron filings are removed and new ones are placed on the paper, they arrange themselves in the same pattern as the previous ones: the energy field of the magnet creates the pattern. Electromagnetic fields exist in nature, one of them being the Earth's magnetic field, which can be observed on a compass. Other celestial bodies like stars and planets also have electromagnetic fields, and so do smaller objects like rocks and crystals.

Electromagnetic fields have also been measured extending around humans, animals, and plants. What's most remarkable is that bees, algae, worms, ants, insects, hummingbirds, bumblebees, and dolphins, among others, all have the ability to perceive these electromagnetic fields.[15] In other words, fields are "energetic blueprints" or templates of organisms. For example, the design of a rose—which includes color, shape, scent, structure, and genetics—first exists in the field of the flower.

Dawson Church in his book *Mind to Matter: The Astonishing Science of How Your Brain Creates Material Reality*, cites a study published in the journal *Science* in which researchers reported that bees have the ability to detect the fields around flowers and use the information to determine which ones have the most nectar. The electromagnetic properties of fields around flowers, Church notes, "came as a surprise to scientists immersed in matter-bound explanations." He adds that after reading the article, Thomas Seeley, a behavioral biologist at Cornell University, said, "We had no idea that this sense even existed."[16]

Knowledge about the fields that encircle living organisms, as documented in contemporary scientific research, may come as a surprise to some, but they have been known of for at least a few decades. Harold Saxton Burr was a professor at Yale School of Medicine in 1929 and studied energy fields around animals and plants, measuring ways in which matter (atoms, molecules, and cells) is organized by those fields as the organisms develop. In 1949 he published his findings of the electromagnetic field around a single nerve, similar to electromagnetic energy that creates patterns on a sheet of paper sprinkled with iron filings. Burr found that fields weren't just produced by living organisms, but that fields created matter, providing lines of force around which matter could arrange itself into atoms, molecules, and cells.[17]

Church quotes Burr : "Something like this [analogous to iron filings on a sheet forming patterns due to electromagnetism] ... happens in the human body. Its molecules and cells are constantly being torn apart and rebuilt with fresh material from the food we eat. But, thanks to the controlling [life]-field, the new molecules and cells are rebuilt as before and arrange themselves in the same pattern as the old ones."[18]

Reviewing Burr's research of how fields create matter, Church explains, "When you cut your finger and your skin regrows, the field provides the blueprint around which the new cells organize themselves. Energy is not an epiphenomenon of matter; energy is organizing matter." Church discusses the research Burr did with salamanders, finding that the voltages on the outer membranes of salamander eggs were quite high on one spot, while on a spot 180 degrees opposite they were minimal. When salamanders grew to maturity, Barr found that the point with the strongest field in the egg became the head and the point with lowest electrical activity would always be the tail. The field, it seemed, had arranged the heads and tails of salamanders during gestation and development.

Church goes on to explain that Burr used mice to determine whether the energy field played a role in cancer. After measuring their fields, Burr noted which mice later developed cancer. Based upon "10,000 measurements, he found that the electromagnetic signature of cancer appeared in the mouse's energy field before any detectable cellular malignancy was evident."[19]

Mind to Matter

Astronaut Edger Mitchell walked on the surface of the moon on February 3, 1971. As he journeyed home, while silently observing the heavens from the space capsule's window, he had a transcendent experience in which he felt one with the entire universe. This experience changed his life: he realized that the nature of the universe was not as he had been taught. This startling realization by a man who was a navy pilot, quantum physicist, and astronaut grabbed my attention! I wanted to know more, and found plenty of answers in his 1996 book, *The Way of the Explorer: An Apollo Astronaut's Journey Through the Material and Mystical World*. In it, among other things, Mitchell documents personal observations of psi phenomenon and mind-matter interactions.

In the fall of 1972, Mitchell went to Little Rock, Arkansas, to speak at a convention about his trip to the moon and his transcendent experience in space. It was his first presentation, and his mother from Oklahoma was in attendance. Her eyesight at the time was extremely poor as a result of glaucoma. In spite of being legally blind without her eyeglasses, she had ruled out corrective surgery because she felt it was too risky.

During the conference Mitchell met several remarkable men and women, among them an American named Norbu Chen who claimed to be a healer and had studied the earliest form of Tibetan Buddhism as well as shamanistic practices. Though Mitchell was skeptical about Chen's healing abilities, he was also curious. So, one evening during the conference, he introduced Chen to his mother, who was at the time in her early sixties, to see if he could help her.

As a Christian fundamentalist, Mitchell's mother believed the mind could influence matter only through either God or Satan. Chen did not think that way but was sure he could help, and she agreed to let him.

The next day Chen came to help her in Mitchell's suite. After getting into a meditative trance and chanting a mantra, he floated his hands over her eyes. Mitchell observed his mother demonstrating a calm and silent trust in this man whom she had never met before. After a few minutes, Chen gently told her that he was finished and she should rest and treat herself kindly.

Mitchell was both skeptical and hopeful for a positive outcome. The next morning at six o'clock, Mitchell's mother rushed into his suite and excitedly told him that she could see. She dropped her thick glasses to the floor and crushed them with her heel, saying softly, "I can see. Praise the Lord, I can see."

Mitchell was both pleased and bewildered. He noted in his book, "This wasn't science, but as far as I was concerned, it indicated where I personally needed to probe more thoroughly. All I can say is that it absolutely did happen in just this way. Afterward I experienced the deep-down astonishment that arises from witnessing the extraordinary. This was an event I couldn't explain, but I couldn't deny it either. I knew my mother's reaction was authentic, and she hadn't been duped about her own sight. She proceeded to drive home alone, several hundred miles, without her glasses. After this episode I was sufficiently impressed, so I invited Norbu to Houston for a visit so that I might learn a few things from him myself."[20]

Chen struck Mitchell as a rather ordinary person who nonetheless had a remarkable capacity to heal that Mitchell couldn't adequately explain. He was left wondering how Chen had used his mind to make such a dramatic physical change in his mother's eyes.

There are thousands of healers around the world who, like Chen, cannot offer an adequate explanation of their mental capacity to interact with matter. But neither are there any adequate scientific explanations for the placebo effect, nor do we know how spontaneous healing, distant healing, or any psi phenomenon actually works. As we've discussed in previous chapters, nonordinary phenomena are intriguing and mystifying only because our understanding of the nature of reality is inadequate. To gain a broader and deeper understanding, let's look further into Mitchell's experiences with psychics like Chen and Uri Geller.

After Mitchell's mother returned home, she was able to go about her daily activities without wearing glasses or contact lenses—her vision remained nearly perfect. Then one day she called Mitchell to ask him if Chen was a Christian. Mitchell was reluctant to reveal that Chen was not, but she insisted on knowing. When she learned that Chen was of a different faith, with deep regret she concluded that the healing could not have been the work of the Lord after all, but of darker, satanic forces. Mitchell tried to explain her recovery in secular terms, but his mother's mind was made up. Mere hours later, she became nearly blind once again.

Mitchell was dismayed and confused. In spite of his mother's experience and the extraordinary healing he had witnessed, he still couldn't accept that beliefs had the power to affect matter. He admitted, "For several years I would continue to underestimate the power of belief in our lives because of the pervasiveness of my classical scientific training. It still puzzled me that belief could affect anything at all."[21]

But Mitchell remained curious. He sought out many other psychics in order to gain further evidence and understanding concerning Norbu Chen's healing abilities. Along the way, he encountered many remarkable healers, along with some fraudulent ones. One of the most gifted people he met was the Israeli psychic Uri Geller, whose mental ability to interact with material objects was studied under controlled scientific protocol by Mitchell and colleagues at the Stanford Research Institute.

Mind-Matter Interactions

Mitchell met the then twenty-five-year-old Geller in 1972. At the time, he was not well known. Dr. Andrija Puharich, an American physician who spent months observing Geller's extraordinary abilities in telepathy, clairvoyance, and psychokinesis, wanted to test them in an American laboratory. He contacted Mitchell, who arranged for Geller to come to SRI for these tests. A team of scientists that included physicists Hal Puthoff and Russell Targ performed double-blind experiments on Geller to determine the extent of his psi capabilities.

Geller was an ordinary youngster with an air of flamboyance and showmanship. In his childhood, he had considered normal what others felt were extraordinary psychic powers. As an adult he performed jaw-dropping acts on television, which most people thought were simply staged magic tricks. But as Mitchell and the scientists at SRI discovered, Geller's "magic tricks" were mostly demonstrations of his genuine psychic talents.

The first double-blind studies were designed to test Geller's ability to view objects at a distance, as had been done with Ingo Swann and Hella Hammid, discussed in chapter 3. Geller sat in a Faraday cage, which blocks electromagnetic waves and thus prevents all forms of electronic communication. He was given targets chosen at random to describe. Almost every time, Geller was able to draw the target sites fairly accurately.

One of Geller's signature capacities was his ability to bend metal objects like spoons and forks using the power of his mind; his gently touching the spoon with one finger was enough to bend it into a coil. But scientists who were invited to witness these experiments believed Geller was a fraud, claiming he was using some kind of special solvent on his finger that caused the spoon to bend. The physicians in the group were at a loss to explain how the mere touch of a finger could bend a metal object like a spoon. Though no solvent was used, and no force beyond a simple touch was applied, the physicians continued to deny what they themselves had witnessed.

Mitchell's most astonishing experience was seeing children bend forks and spoons using their thoughts after doing nothing more than watching Geller on television! He explained:

> I went to a number of homes around the country, sometimes with my own spoons in pocket, or I would select one at random from the family kitchen. Typically it was a boy less than ten years of age who would lightly stroke the metal object at the narrow point of the handle while I held it between thumb and forefinger at the end of the handle. The spoon would soon slowly bend, creating two 360-degree twists in the handle, perfectly emulating what Uri demonstrated on television. No tricks, no magic potions, just innocent children (with normal children's fingers) who had not yet learned that it could

not be done. The evidence continued to mount in this way, suggesting that these strange capabilities were quite natural and likely common in humans, though latent and seldom manifest.[22]

Mitchell reported that Professor John Hasted, chairman of the Department of Physics at Birkbeck College in London, and physicist Ted Bastin both conducted extensive experiments with children and found that many of them could bend metal objects without any physical contact with them. Mitchell noted that acclaimed physicists from the United States, Britain, Denmark, France, and Germany also successfully tested Geller and/or the "Geller children." In 1977 they gathered in Iceland to report their results and theories. Their findings were not accepted in professional journals, however, so they documented their results in *The Iceland Papers*, with a foreword by Nobel physicist Brian Josephson, which they published themselves.[23]

In the modern era, the gold standard of scientific evidence is to get published in peer-reviewed journals. It is important to note, however, that when research carried out by noted scientists and supported by a Nobel physicist is not published because the results do not fit neatly into the existing scientific paradigm, the problem is not the validity of the studies or their conclusions. What needs be questioned are the culture and mind-set of the scientific community itself.

Geller's mind-matter interactions didn't end with bending metal objects. During the six weeks Mitchel performed experiments with Geller at SRI, there were numerous incidents of electronic equipment failures no one could explain. Video equipment to which Geller had no access would suddenly lose a pulley, which would be found later in an adjoining room. Jewelry would go missing, only to be found in a locked safe with a combination unknown to Geller. In spite of equipment failures and objects moving around for unknown reasons, Mitchell and the other scientists carried out several well-designed experiments to test Geller's psychokinetic capabilities.

In one experiment Geller was asked to move a large ball bearing under a glass jar without touching it or the table it was on. It took him several minutes, but he did it, and the event was recorded on video. According to Mitchell:

> Finally, we had something legitimate on tape. When we went to review what we had all seen in person, we were relieved to see that the event was caught cleanly within the camera's field of view. Nothing had gone amiss. However, the feat was still greeted with skepticism when our colleagues in science viewed what everyone who had witnessed the event and had thought monumental. They became red in the face, and some left, refusing to ever return to the lab. They accused Uri of being a fraud, and the rest of us of being chumps in an elaborate charade. But their accusations flew in the face of the solid scientific work that had been done, and I believe they knew it. Even some visiting scientists who watched positive results directly and in person angrily rejected what they saw.[24]

Denial of factual information and events that defy common sense and the laws of classical physics is not limited to scientists. Most people find psi phenomena hard to believe, but those who have had personal experiences that can't be explained find it reassuring to learn what scientists like Mitchell have carried out.

For me personally, some of the most intriguing and mind-altering events Mitchell reported included Geller's telekinetic abilities (telekinesis means transporting a material object by mental means). To test this ability, Mitchell asked no less than for Geller to recover the camera he had left behind on the moon.

This request spurred a series of strange events. A few days later as Geller was having ice cream at the SRI cafeteria during lunch, he cried out in pain as blood seeped from his lips. He had bitten into a tiny piece of metal. As seven or eight others watched, he pulled the object from his mouth, washed it in a glass of water, and handed it over to Mitchell. It was a piece of tie clasp given to Mitchell a few years earlier that he had lost along with a box containing cuff links and tie pins. People at the lunch table chuckled nervously as they witnessed this strange occurrence.

As Mitchell was walking back to work afterward, he heard metal strike the tile floor. He turned around to see Dr. Puthoff picking up a small, shiny object from the floor. Not knowing what it was, Puthoff handed it to Mitchell, who recognized it as the other half of the broken tie clasp that had appeared moments earlier in Geller's mouth. Mitchell noted that the atmosphere was growing downright eerie, and he and his colleague laughed nervously as they returned to work.

Moments later Mitchell saw something fall on the floor between himself and Puthoff. Bewildered, he reached down to pick it up: it was a tie pin his brother had given to him years earlier, which he had kept in the jewelry box he lost. These three telekinetic events took place within the span of thirty minutes in full view of some of the best scientists of the time.

When I first read about mind influencing matter in Mitchell's book, I was amazed. I wondered how such events could happen, and it was hard for me to understand how it was possible. Sometime later, I experienced an incident that made it impossible for me to ignore telekinesis and the reality of the mind-matter connection.

Was It Telekinesis?

One morning in 2008, two years before my father died, he was living in Lahore and in poor health. My sister and her husband, with whom he lived, suddenly noticed that Dad was unable to talk or respond to their touch. Realizing he needed immediate medical attention, they called for an ambulance. On the ride to the hospital, he flatlined—instruments registered no heartbeat. While he was being transported to the hospital, my sister called me and said, "It's up to you, but you should make plans to come to Lahore, because Abbu (Dad) may not live."

I had mentally prepared to get such a phone call, since

Abbu had been quite sick, and I anxiously awaited a follow-up report on his condition. An hour or so later, my brother-in-law called from the hospital and said, "Abbu is in stable condition." I was relieved to hear this news, but we all knew that Dad's "stable condition" was precarious. I prayed for his well-being and went to sleep that night knowing he was being cared for and hoping he was comfortable. He responded well to emergency medical procedures, and a couple of days later was well enough to return home.

The next morning I went to the guestroom in my house where my dad stayed when he came for extended visits. As I walked into the room I saw that the glass dome covering the light fixture on the ceiling had fallen onto the bed directly underneath it. I thought, *Wow! It's a good thing that the lamp didn't fall on Abbu when he was here.*

I climbed onto the bed to reinstall the glass dome—and was instantly confused. The three metallic screws that held it in place were fully tightened in their normal positions and should have held the dome fast. I had assumed they would be so loose that I would simply have to insert the lamp and tighten them. How could the glass dome fall when the screws holding it were in the correct positions? And, yet it had!

As I wondered how such a thing could have happened, I intuitively felt that it was my father trying to reach me telekinetically, perhaps when he flatlined on the way to the hospital. In excited amazement, I called to my girlfriend, who was downstairs, to come up and look at the scene on the bed. I felt a bit strange as I nervously said, "I think it was

Above: Two of the three screws on the light fixture are visible in their fully tightened position. The third fully tightened screw is not visible in this picture.

Below: The glass dome exactly where it fell.

Abbu trying to reach me." She not only agreed, but encouraged me to take some pictures, which I did. Had this event really been the result of my dad's desire to communicate with me while being transported in the ambulance? In my heart I felt that it was.

* * *

The validity of psi phenomena documented by such credible sources as Mitchell and many others, along with my personal experiences, led me to realize that we live in an infinitely magical universe. And these events guided me toward a greater awareness of who we are and what we are capable off. As you continue to read this book, I believe that you, too, may experience a shift in your understanding of the nature of reality. It may not happen right away; after all, at first, I was in a "Gee-whiz" state of mind myself. But gradually, as you reflect on and contemplate what you are learning, and especially when you experience psi phenomenon yourself—which most commonly take the form of coincidences and dreams—you will gain deeper acceptance and understanding of the insights and information I am presenting in this book.

You may start to sense the presence of all physical phenomena in their energetic and physical form simultaneously. You may start to see and feel that you are not just a physical or biological being, but an energetic presence with unique vibrational patterns. You may find that you possess gifts you never thought you had or could possibly have, gifts that can help you attain greater health, well-being, and success in all aspects of your life. In the next chapter, we will review findings in quantum physics and the scientific evidence of how mind and matter interact, as well as other details regarding our "magical" universe.

5

Consciousness—
The True Nature of Reality

I regard consciousness as fundamental. I regard matter as a derivative from consciousness. We cannot get behind consciousness. Everything that we talk about, everything we regard as existing, postulates consciousness.
—Max Planck, Nobel laureate in physics, 1918

*The fact that consciousness is inseparable
from cognition, perception, observation, and
measurement is undeniable; therefore, this
is the starting point for new insights into the
nature of reality.*
—Menas Kafatos, Rudolph Tanzi, and
    Deepak Chopra, Cosmology of
    Consciousness: Quantum Physics
    and the Neuroscience of Mind

*The emerging understanding of reality intro-
duced by quantum theory and the gathering
evidence of the active role of consciousness
in the world and the cosmos not only bring
about a basic paradigm shift, but give us the
basis for a whole new story that is at once
very ancient and ever new.*
—Jean Houston, What Is Consciousness?

I n the previous four chapters, we discussed the fact that
things may not be as we have assumed them to be. In
this chapter we will continue to contrast our normal
understandings about life and the universe with how things
actually are. To align our thinking with the true nature
of reality, we will examine consciousness and reflect on
its being fundamental to the existence of all phenomena,
tangible and intangible.

Consciousness can be understood in two ways. Normally when we use the word, it refers to being conscious,

aware, awake, or observant—or the opposite: unconscious. In *metaphysical idealism*, as well as science and ancient spiritual understandings, it refers to the universal mind, which is omniscient, omnipresent, and beyond space-time. This second, more expansive meaning of consciousness will be the focus of discussion in this chapter.

To develop the understanding that consciousness is the true nature of reality, we will review some important established scientific facts that, upon further inquiry, need reinterpretation. The process of arriving at different understandings based upon new discoveries is the hallmark of science. It has helped us gain greater knowledge of life and the universe. Similarly, reviewing the latest scientific information about how life originated in the stars, new facts about the Big Bang, salient features of quantum theory, and recent findings in biology and neuroscience may guide us to see that consciousness is the source of all phenomena in the cosmos. In reinterpreting how things really are, we recognize our potential to access the power of consciousness or infinite mind, which can help us achieve a prosperous, joyous, and meaningful life.

* * *

Note: I suggest that before you read this chapter, you review appendix A on page 293, a summary of the topics covered here and discussed throughout the book. It contrasts conventional thinking about these topics with emergent thinking, and you may find it a useful overview.

Quantum Theory and the Nature of Reality

For most of us, the intricate details of the mathematics and physics of quantum theory are difficult to understand, and the universe in its infinite complexity is unknowable. However, because consciousness is fundamental, human beings are born with a sense of deeper knowing about everything, including the salient features of quantum theory—though we may not be consciously aware of them. Understanding consciousness as the true nature of reality permits us to make sense of the NDEs, intelligence in nature, universal mind, and mind-matter connections we discussed earlier. To explore consciousness from a scientific perspective, we begin with the wave-particle paradox, entanglement, and complementarity in quantum theory. Some of the following discussion may seem difficult to follow; however, most of the chapter will not be. I have tried to follow Einstein's advice: "Everything should be as simple as it can be, but not simpler."

Approximately twenty-five hundred years ago, the sages in India explained that physical reality was an illusion. Furthermore, it was noted in the *Yoga Sutras of Patanjali* that everything in the universe was interconnected, interdependent, existed as one reality, and contained all information and knowledge, which could be accessed.

Interestingly, in the earlier part of the twentieth century, theoretical physicists began questioning the prevailing understanding of the nature of reality because their stunning discoveries required them to reevaluate classical physics, according to which matter was considered to exist independently. Their new findings suggested that matter was not independent; it came into existence only when observed. This extraordinary, counterintuitive knowledge about the nature of subatomic particles, and the consensus about it among the physicists of the day, gave birth to the quantum revolution. It brought forth the realization among the most brilliant minds in physics, more than a hundred years ago, that consciousness is foundational to all phenomena of the known and unknown worlds.

However, not all physicists of the earlier twentieth century interpreted the new discoveries the same way. Manjit Kumar, in his book, *Quantum: Einstein, Bohr and the Great Debate About the Nature of Reality*, writes that a week-long conference called "Electrons and Photons" was held in Brussels in October 1927. No fewer than seventeen of the twenty-nine scientists invited to the gathering were then or eventually became Nobel laureates in physics. There, they engaged in a debate about their spectacular findings, the very nature of reality, and the soul of physics. One of the most spectacular meetings of minds ever held, it marked the end of the golden age of physics, an era of scientific creativity unparalleled since the scientific revolution in the seventeenth century.[1]

What were these findings? And who won the debate? The title of Kumar's book tells us that two geniuses, one now better known than the other, Albert Einstein and Niels Bohr, took opposite sides in the great debate. The other physicists present, to varying degrees, sided with either Einstein or Bohr. The grand debate could have produced a clear winner, but this didn't happen, and in many ways the same debate about quantum theory continues to this day. Subsequent findings have refined and added to the points of view Bohr had expressed. We will explore the salient features of these differing perspectives, and subsequent refinements, to gain perspective about quantum theory.

In 1935 Einstein and his colleagues Boris Poldolsky and Nathan Rosen introduced the concept of *entanglement* in a thought experiment that suggested subatomic particles, once brought into close proximity, would maintain their connection regardless of the distance between them, without expenditure of time or energy. This came to be known as the EPR (Einstein, Poldolsky, and Rosen) experiment. However, Einstein thought the strange connection between subatomic particles that had been separated from one another was not possible in actuality. In fact, he referred to it as "spooky action at a distance." Why did Einstein refuse to accept his own finding? Perhaps because of his groundbreaking and acclaimed general theory of relativity, published in 1905, in which he proved that the laws of physics are the same for all nonaccelerating observers, and that the speed of light (186,000 miles per second) is a cosmic constant; nothing could travel faster. By contrast, the EPR experiment suggested that instantaneous corresponding action between subatomic particles meant they communicated with one another faster than the speed of light. Was this Einstein's concern when he debated Niels Bohr? It may have been.

Another fundamental concept of quantum theory that Einstein didn't accept was the *wave-particle duality*. The double-slit experiment, which was carried out to determine whether light was a particle or a wave, had shown that it could be either one or both at once, and that its nature was dependent on the observer. This wave-particle duality is at the heart of quantum theory, which was Bohr's point of view. Einstein did not accept that thoughts or observation affected the behavior of subatomic particles.

The Nonduality of Mind and Matter

What did the double-slit experiment show? That thoughts create the physical world! This notion so radically differs from the manner in which our senses perceive the nature of reality that the evidence provided in the experiment, conducted almost a hundred years ago, is still vigorously debated. Though the vast majority of physicists accept that in the subatomic world the observer effect is real, many reject that thoughts create matter in the macro world of sticks and stones. Einstein quipped, "If I'm not looking at the moon I would like to believe that it's still there."

How should we interpret his statement? Einstein's comment regarding the everyday physical world was simply an obvious statement laced with gentle sarcasm. However, the wave-particle duality implies that everything exists as infinite possibilities until observed. Therefore, in the context of the observer effect, his statement wouldn't be correct. But Einstein being Einstein, he ought not to be dismissed too quickly. And indeed, we will return to this issue.

The double-slit experiment has been repeated thousands of times by scientists from all over the world, and the same results have been observed. In fact, today this experiment appears in high school physics books. An animated version of the double-slit experiment demonstrated by "Dr. Quantum" in the movie *What the Bleep Do We Know!?* is well done and makes understanding the experiment easier. You can find it on YouTube.

The Quantum Debate Continues

Niels Bohr (1885–1962) was a Danish physicist and philosopher who was awarded the Nobel Prize in Physics in 1922. His work was foundational to the development of quantum theory, and according to the great physicists of his time—some of whom were present at the conference in Brussels—he was considered intellectually superior to Einstein. Contemporary quantum physicists also consider Bohr's contributions to be monumental, for he postulated that microphysical objects have no intrinsic properties:

> An electron simply does not exist at any place until an observation or measurement is performed to locate it. It does not have velocity or any other physical attribute until it is measured. In between measurements it is meaningless to ask what is the position or velocity of an electron. Since quantum mechanics says nothing about a physical reality that exists independently of the measuring equipment, only in the act of measurement does the electron become "real." An unobserved electron does not exist.[2]

Bohr contradicted the assumption in classical physics about the independent nature of a subatomic particle. Instead Bohr's *complementarity principle* postulated that a complete knowledge of phenomena on atomic dimensions requires a description of both wave and particle properties. It is impossible to observe both the wave and particle aspects simultaneously.[3]

In1905 Einstein showed that energy and matter are one and the same with his famous equation $E = MC^2$. In the realm of the physical or the *local domain*, Einstein's brilliant and most elegant formula brought scientific understanding of the nature of reality a step closer. But it took several other physicists in the early part of the twentieth century— beginning with Max Planck in 1900, who is regarded as the originator of quantum theory, along with Niels Bohr, Paul Dirac, Werner Heisenberg, John von Neumann, Erwin Schrödinger, and many other physicists who were present at the conference in Brussels—to conclusively arrive at an interpretation of quantum theory that included the role of *thought* or *consciousness* as foundational to understanding the nature of reality. Sir James Jeans, whose contributions to quantum theory were considerable, concluded:

> The stream of knowledge is heading towards a non-mechanical reality; the universe begins to look more like a great thought than like a great machine. Mind no longer appears to be an accidental intruder into the realm of matter ... we ought rather hail it as the creator and governor of the realm of matter.[4]

Einstein's version of the nature of reality was interpreted within the parameters of the physical universe, and he postulated that the universe was made up of only *energy* and *matter*. However, quantum theory asserted that thought or consciousness was foundational. This was at the heart of why Einstein disagreed with Bohr and declared quantum theory to be "incomplete."

Who won the debate? Those who were in Bohr's camp regarding quantum theory maintained their support. John Archibald Wheeler in particular was emphatic: "Bohr's principle of complementarity is the most revolutionary scientific concept of this century and the heart of his fifty-year search for the full significance of the quantum idea."[5] The complementarity principle, wave-particle paradox, and entanglement have all stood the test of time, and today these concepts of quantum theory are regarded as fundamental. Einstein eventually agreed with Bohr and the fundamental principles of quantum theory, but he maintained his doubts.

The Convergence of the Ancient and Modern

In the Buddhist philosophy of *emptiness*, there is a deep recognition that the way we perceive the world is not in accordance with the way the world actually is. On a day-to-day basis, our senses inform us of a discrete, independent, and definable reality in which everything exists, whether anyone is there to observe it or not. For example, the sun, moon, stars, and everything in the universe are present irrespective of anyone's observation. However, according to the Dalai Lama:

> Any belief in an objective reality grounded in the assumption of intrinsic, independent existence is untenable. All things and events, whether material, mental, or even abstract concepts like time, are devoid of objective, independent existence.[6]

He explains that the fundamental truth of "the way things really are" in Buddhism is referred to as *shunyata* or emptiness, and adds, "Things and events are 'empty' in that they do not possess any immutable essence, intrinsic reality, or absolute 'being' that affords independence."[7]

Quantum theory's findings of the complementarity principle, wave-particle duality, entanglement, and non-locality also suggest that the universe is not an independent, definable, discrete, or enduring reality. Rather, it is dependent on the thoughts, feelings, and observations of the observer. In other words, quantum theory also implies emptiness, oneness, intelligence, and an intrinsic mystery about the universe. Astrophysicist Sir Arthur Eddington (1882–1944) mused, "Something unknown is doing, we don't know what."[8] Thus concepts of quantum theory and Buddhism's philosophy of emptiness have unmistakable resonance. According to the Dalai Lama, "Science is coming closer to the Buddhist contemplative insights of emptiness and interdependence."[9]

As I mentioned earlier, the idea that infinite mind or

consciousness is foundational, and that it is the source of all things in the world, has been around for a long time. In Western thought the philosophy of metaphysical idealism expresses it best. Philosopher and scientist Bernardo Kastrup, in his book *The Idea of the World: A Multi-Disciplinary Argument for the Mental Nature of Reality*, writes, "A universal phenomenal consciousness is the sole ontological primitive, whose patterns of excitation constitute existence.... The inanimate universe we see around us is the extrinsic appearance of a possibly instinctual but certainly elaborate universal *thought*, much like a living brain is the extrinsic appearance of a person's conscious life."[10] He defends his assertion by presenting evidence from such diverse fields as philosophy, neuroscience, psychology, psychiatry, and physics. His argument that the nature of reality is mental when seen collectively from various perspectives leaves the philosophy of scientific materialism or physicalism untenable. However, many mainstream scientists remain unconvinced.

The Hard Problem of Consciousness

In 2014 I attended the "Toward a Science of Consciousness" conference in Tucson, Arizona. Stuart Hameroff, an anesthesiologist and consciousness researcher at the University of Arizona, was the conference organizer. Featured speakers included Oxford mathematician and physicist Sir Roger Penrose, Deepak Chopra, MD, and others. I was interested in attending this conference because I was in the midst of my research for this book, and keen on hearing a few scientific luminaries of our time in person. In the movie *What the Bleep Do We Know!?* Hameroff asks, "Where does consciousness go when someone is under anesthesia, and how does it return?"—a question that made me curious. Australian philosopher David Chalmers, in his 1996 book *The Conscious Mind*, questioned how and why subjective mind appears to arise from objective mind. In other words, how do the feelings of awe, wonder, joy, love, creativity, sadness, or apathy seem to arise from matter (the brain)? There were no reasonable answers in any of the sciences, which is why Chalmers coined the term "the hard problem." Hameroff and Penrose were both interested in finding answers to this difficult question.

Hameroff's work on very small sections within human brain cells called microtubules had caused him to suspect that the latent field of consciousness embedded in space-time may be functioning in these tiny structures of the brain. Penrose's work in quantum physics, gravity, and the geometry of space-time was immense, and because he had decided to work with Hameroff, I wanted to know what they were up to and why.

On the first day of the conference it was obvious that most of the presenters and attendees were scientific materialists. Since an objective of the conference was to explore the "hard problem" of consciousness, several of the speakers made fleeting reference to this issue. With the exception of Deepak Chopra, however, the dominant perspective of the speakers was that consciousness is created by the brain. There were variations in their perspectives. One was to simply ignore the "hard problem." In the words of one presenter, "It's a waste of time. Scientists and engineers should be doing research, making better computers, cell phones, or other useful devices." Another perspective was straightforward: "Evidence that brain creates consciousness will be discovered in the future," although no specific reasons or scientific findings were discussed to back this assertion. A third perspective was rather stunning: "Consciousness is an illusion created by the biochemical reactions in the brain. There is no such thing as consciousness." This computer scientist explained that it would be possible to develop very high speed computers that would be creative and solve problems much better and faster than human beings can. When asked how it would be possible for computers to think, express emotions, and be creative just like humans, only better, he pointed to the massive memory of the superfast computers of the near future. Amazingly, this comment drew enthusiastic applause, and, not for the first time, I was impressed by the power of dogma. In fact, the firsthand experience of witnessing the entrenched ideology of scientific materialism turned out to be my most valuable takeaway from the conference. It helped me realize that the tone and focus of the book you are reading ought to be geared toward those who are open to new ideas.

Hameroff's and Penrose's presentations outlined how it might be possible for consciousness, which is embedded in space-time, to present itself in the brain's microtubules. It was inspiring and endearing to see the soft-spoken English gentleman Sir Roger Penrose's presentation, which he delivered with humility and grace. He used an overhead projector and a magic marker to highlight his ideas, drawing an audible gasp of disbelief from a younger audience used to elaborate PowerPoint presentations including multicolored graphics and sound. Hameroff's arguments for a relationship between quantum theory and consciousness were interesting and persuasive, although, as he acknowledged, his theory was yet to be proven.

In essence, Hameroff's and Penrose's theory is that consciousness is woven into the fabric of space-time itself, and the coherent quantum activity among the microtubules in the brain seems to create our consciousness. Furthermore, this theory states that there is one common underlying entity that gives rise to, on the one hand, matter, and on the other hand, mind, and this single entity is quantum space-time geometry. Though these arguments are compelling, there are some fundamental problems with them. For example, while microtubules are present in the brain, how does consciousness function in a plant, a tree, or a drop of water, none of which has a brain?

Hameroff and Penrose acknowledged the presence of consciousness as fundamental, but in their efforts to explain how it functions in the brain, they suggested materialistic inclinations. My understanding, on the other hand, is in alignment with *metaphysical idealism*, which asserts that mind or consciousness is fundamental. My decades-long research in various fields of study concerning consciousness has informed me that all phenomena including quantum space-time geometry, matter, and mind are representations of consciousness, and not limited to the confines of a brain. Near-death experiences, out-of-body experiences, and extrasensory perception all indicate that consciousness is fundamental, underlying all phenomena.

Perhaps in the future Penrose and Hameroff will shed more light on how consciousness and matter (brain) interact, but I don't think the infinite beauty, complexity, and mystery of consciousness or the nature of reality can be explained in the rational, observable, measurable, or logical framework of scientific materialism. The "hard problem" exists only for the physicalists. It doesn't exist when consciousness is considered to be the true nature of reality. We can only experience consciousness. It cannot be understood within the illusions of reality created by our senses.

Vibrational Energies of Consciousness and the Illusion of Reality

As I touched on earlier, according to ancient Indian philosophy, physical reality is an illusion or *maya*, and in essence, this notion is also at the heart of quantum theory. But then why is it that our experiences in life, and the physical realities of sticks and stones, planets and solar systems, galaxies and the universe, do not appear to be illusions? How are we to conceptualize the illusion of physical reality? To answer these questions, it is helpful to see how the brain and our five senses in concert create the physical realities we experience.

Sight, sound, touch, smell, and taste are interpretations of the vibrational energies of the atoms and molecules that are all around us and everywhere in the universe. Our five senses do not represent actual reality, nor are their interpretations universal among the various species on earth. For example, our experiences and sensations are different from how dogs, dolphins, whales, birds, bees, bats, and insects experience the same physical world we all inhabit.

Ervin Laszlo, PhD, philosopher and systems scientist, explains that *all phenomena in the universe exist as the vibrational energy of consciousness.* Physical objects, in reality, are just clusters of energy vibrating at different levels. So are nonphysical phenomena like light, sound, and heat. Vibrational energy patterns are received by sensory organs (such as eyes and ears) and are conveyed as electrical impulses to the brain, where they are decoded and give rise to the sensations of color, sound, texture, odor, and taste.[11]

Expressing how our senses create the notions of reality we experience, and in agreement with Laszlo, neuroanatomist Jill Bolte Taylor, PhD, of Harvard University states:

> Our visual field, the entire view of what we can see when we look out into the world, is divided into billions of tiny spots or pixels. Each pixel is filled with atoms and molecules that are in vibration. The retinal cells in the back of our eyes detect the movement of those atomic particles. Atoms vibrating at different frequencies emit different wavelengths of energy, and this information is eventually coded as different colors by the visual cortex in the occipital region of our brain.[12]

Human beings are limited to seeing light within the frequencies of 430 to 770 T (trillion) Hz. The infrared frequency of light is below 430 THz, making it invisible to us. However, mosquitoes, vampire bats, bedbugs, beetle species, and some snakes can see in the infrared spectrum, allowing them to realize a different aspect of physical reality. We cannot see ultraviolet light because its frequency is greater than 770 THz. But animals such as birds, bees, and certain fish can perceive ultraviolet light, and therefore, they see a totally different world.[13] Light frequencies around 430 THz create the perception of red, and frequencies near 770 THz appear blue. In actuality there is no such thing as color.[14]

After I wrote the last paragraph I decided to take a stretch break and walked to the window of my bedroom through which I have a lovely view of my backyard. There are dozens of trees and a wetland adjacent to my property that attract a lot of birds, especially in the spring. Just as I approached the window, I saw a blue jay perched on the nearest branch. I knew that blue jays in actuality are not blue—they are brownish gray. Their feathers have very tiny ripples or indentations, which makes the light reflect off them such that we perceive them to be blue. The coincidence of seeing a blue jay at the exact moment I was writing about the illusory nature of colors was humbling. It reminded me that consciousness is at the heart of all of our thoughts, feelings, experiences, and the oneness of our existence. The illusory nature of light is an example of how our brains interpret reality rather than perceiving its true nature.

Says Jill Bolte Taylor:

Similar to vision, our ability to hear sound also depends upon our detection of energy traveling at different wavelengths. Sound is a product of atomic particles in space colliding with one another and emitting patterns of energy. The energy wavelengths, created by the bombarding particles, beat upon our tympanic membrane in our ear. Different wavelengths of sound vibrate our eardrum with unique properties. Similar to our retinal cells, the hair cells of our auditory Organ of Corti translate this energy vibration in our ear into a neural code. This eventually reaches the auditory cortex (in the temporal region of our brain) and we hear sound.[15]

A young, healthy person can hear sound frequencies in the range of 20 to 20,000 Hertz. In contrast, depending on the species, dogs can hear in the range of 40 to 60,000 Hz. This is why dogs placed behind invisible fences remain there. Sound waves emitted when the dog crosses the invisible barriers are in the higher range of its hearing, which is quite unpleasant for the dog, though humans cannot hear those sounds. Other animals with much greater ability to hear sound include mice (1,000 to 91,000 Hz); bats (2,000 to 110,000 Hz); and beluga whales (1,000 to 123,000 Hz).[16] This illustrates once again that the nature of reality as experienced by different species is dependent on their sense organs' capacity to receive the vibrational energies of consciousness.

Bolte Taylor:

Our most obvious abilities to sense atomic/molecu-
lar information occur through our chemical senses
of smell and taste. Although these receptors are
sensitive to individual electromagnetic particles as
they waft past our nose or titillate our taste buds, we
are all unique in how much stimulation is required
before we can smell or taste something.[17]

For instance, dogs have an incredible sense of smell,
and therefore their perception of the nature of reality is
vastly different from ours.

Similarly, skin, our largest sensory organ, is "stippled
with very specific sensory receptors designed to experience
pressure, vibration, light touch, pain or temperature. These
receptors are precise in the type of stimulation they perceive
such that only cold stimulation can be perceived by cold
sensory receptors and only vibration can be detected by
vibration receptors. Because of this specificity, our skin
is a finally mapped surface of sensory reception."[18] The
differences in how our skin responds to the sensory stimulation of vibrational energies create the perceptions of our
physical world. The true nature of reality for every creature
on Earth remains an illusion. However, human beings and
all other creatures on Earth were meant to experience the
illusion as reality. The physical domain and life on Earth
have purpose and meaning, which we will discuss further
later in this book.

Memory Resides in Consciousness

The dominant perspective among conventional neuroscientists and the public at large is that memory resides in the brain. And the neuroscientific community has been trying to locate a site for the storage of memories in the brain for more than a century, but to no avail. A contemporary explanation is that memories are encoded in the molecular details of synaptic connections of neurons with one another. However, this does not explain how the initial encoding of memory in atoms or molecules, which over the years get replaced numerous times, remains intact.[19]

Eben Alexander, whose NDE we discussed in chapter 1, explains in his new book, *Living in a Mindful Universe: A Neurosurgeon's Journey into the Heart of Consciousness,* that memory does not reside in the brain. He states:

My experience proved especially difficult to understand given my prior notion that memory must somehow be stored in the physical brain. For example, with my brain so damaged, how did my memories of pre-coma knowledge and personal events return in the months after awakening in the ICU? Where did they come from? Was it simply that, as the physical brain recovered, memories stored there were refreshed? Given the severity and duration of my illness, such high-level recovery should have been impossible.[20]

Alexander elaborates:

Small regions of the medial temporal lobes (including the hippocampi) seem to be crucial in the general conversion of short-term to long-term

memory, but that does not seem to be the actual
locale of memory storage. Damage here has no
impact on the retrieval of old memories, only on
the formation of new ones. This evidence supports
the notion of the brain as a receiver or filter for
primordial consciousness, but not the producer of
it nor the location of memory storage."[21]

To emphasize the fact that memory doesn't reside
in the brain, Alexander describes the work of Canadian
neurosurgeon Wilder Penfield, who performed surgical
procedures on the brain while patients were awake because
brain tissue doesn't register pain. Penfield discovered that
memories were elicited when electrical stimulation was
administered to portions of the brain, but he found that
such stimulation points were generally inconsistent over
time intervals between operations, confirming that memory
is not stored in any specific regions of the brain. However,
in the earlier part of his career, since electrical stimulations
were eliciting memories, Penfield thought the brain might
be storing memories. But further investigation led him to
believe that memory wasn't stored in the brain; rather, the
brain's lateral temporal cortex had an *interpretive function*,
suggesting that the brain's function was to decode external
information or stimuli.[22]

In chapter 3, the remarkable similarities of the Jim
twins suggested the possibility of nonlocal, mind-to-mind
communication or "entanglement," as in quantum theory.
Could this type of communication or entanglement occur at
the cellular level? If it did, would it further signify that the

brain decodes and processes nonlocal information, rather than storing it? Rupert Sheldrake, in his book *Morphic Resonance: The Nature of Formative Causation*, points to a study conducted by Miroslav Hill, a French cell biologist in the 1980s who showed that cells seem to influence other, similar cells at a distance. In his experiment, cells exposed to toxins developed resistance, but amazingly their "sister cells," which were separated from them and not exposed to toxins, also developed resistance. Hill concluded there was additional flow of information not mediated by DNA. He suggested that because these were descendants of the same mother cell, they were "entangled" and therefore had developed this resistance. Hill's explanation of entanglement at the cellular level also shed light on the remarkable similarities of the Jim twins. If the brains of the twins were independent entities with localized memories and were not entangled, they wouldn't have been so amazingly similar.[23]

As we discussed earlier, Rupert Sheldrake expressed the idea that the brain decodes information like a television set rather than producing it. The program one watches on TV isn't stored in the set; rather, the set receives the broadcast of electrical signals and decodes these signals to produce the pictures that appear on its screen. Sheldrake's analogy suggests that the brain is a receiver of ideas, inspiration, creativity, and memory, as well as our thoughts and feelings, which originate in consciousness.[24]

Though the analogy of a television is appropriate, I think the analogy of interactions between a personal computer and the Internet better explain the *filtering* process of the brain. Everyone is a representation of consciousness and intricately woven into it. However, we all have individual personalities, talents, and preferences, and therefore our connection or interface with consciousness varies. Some of us are mathematicians, while others are musicians, artists, or writers, or excel in sports. Similarly, we use the Internet differently based on our individual preferences, filtering the vast information available to suit our needs. Our filtering process while using the Internet on our personal computers is deliberate. However, in the interface of our brains with consciousness, the filtering of what and how much we receive happens without conscious decision on our part—though not entirely. Everyone has the capacity to gain greater access to consciousness. Therefore, gaining better health, happiness, and success in life and work—including access to greater knowledge, creativity, inspiration, and peace of mind—are all possible. In essence, gaining greater access to consciousness makes attaining any of our hopes, dreams, and aspirations possible. We will cover a few strategies and practices for how to attain greater access to consciousness in chapter 6. But first let's look at the filter theory emerging today in neuroscience and the philosophy of mind.

Recall in chapter 1 when Alexander was in a coma and experienced gaining access to what he considered to be the infinite mind. In this interface he gained instant knowledge, and answers to anything he wanted to know. Yet when he emerged from his coma, he couldn't understand or even remember the insights he gained during his NDE. He described the problem of not being able to recall what he had learned as the brain's "bottleneck" function. In his conscious state, vast amounts of knowledge and understandings were filtered out, and what he did remember was no more than a trickle of all he had learned.

Alexander's firsthand experience of the filtering process of the brain is not an isolated incident. This phenomenon is known as the "filter theory" in neuroscience and the philosophy of mind. "In filter theory, the physical brain serves as the reducing valve or filter through which universal consciousness, or the collective mind, is allowed into our more restricted human perception of the world around us."[25] In other words, the very function of the brain is to create an illusion of physical reality, and it does this extremely well.

Why does the brain create an illusion of physical reality? The simple answer is that we were meant to experience life and all of physical reality just as it seems to be. However, simultaneously we were meant to know that physical reality is an illusion. In this knowing we can make choices and learn to use the power of the infinite mind to live in accordance with our most cherished hopes and dreams. In fact, this message is at the heart of the ancient wisdom traditions, which say that the purpose of life is to awaken. Because life as it appears is like a dream, and we must awaken to experience it in all its richness and magnificence.

Consciousness and the Paradox of Time

In the physical world or the local domain, we experience time as having a past, present, and future. However, according to Einstein, this linear nature of time is an illusion.

He famously said, "People like us, who believe in physics, know that the distinction between past, present and future is only a stubbornly persistent illusion."[26]

Einstein did not reject the existence of time. Instead, he rejected the distinction between past, present, and future. Einstein's conception of time is similar to the philosophy of the Sautrantika school of Buddhism, which also describes time as relative, and suggests that past, present, and future are not separate and independent, but rather interdependent. The Dalai Lama explains, "There is no real time that is somehow the grand vessel in which things and events occur, an absolute that has an existence of its own."[27]

To verify Einstein's theory in which time moves slower for the one who is moving, an experiment was conducted in October 1971 by Joseph C. Hafele, a physicist, and Richard E. Keating, an astronomer, who took four cesium-beam atomic clocks aboard commercial airliners. They flew around the world and compared the clocks against others that remained on the ground at the United States Naval Observatory. When reunited, the three sets of clocks were found to disagree with one another, and their differences were consistent with the predictions of special and general relativity. Though the differences between their times were very small, this simple experiment proved Einstein's theory correct. Similar experiments have been carried out hundreds of times over the years, and the results have always been the same. A connection between space and time has been established, and thus the illusory nature of past, present, and future have become a fundamental understanding in physics.[28]

But why does time seem so real to us? And why are so many observable phenomena intricately linked to directional time? For example, why are the sleep patterns of humans and other species linked to planetary movements called circadian rhythms? Why do some birds lay eggs that hatch synchronous to when caterpillars become available in greater numbers for the birds to feed their young? Why do millions of bats fly out of caves at the precise time when, miles away, they run into swarms of insects coming from the opposite direction, which they then catch and bring back to their offspring? Why does the human body during the "dawn effect" secrete higher levels of adrenaline and growth hormones, along with an increase in blood sugar level right, around 4 a.m.? It is meant to help wake us up and get us ready for the day, but it also signifies time as an integral part of nature. The paradox of the illusion of time as Einstein explained it and our observable connections to time depicted in natural phenomena occurs because human beings experience reality on two planes—the *physical* and the *infinite*—while in the true nature of reality, everything happens in the *present* moment. The concept of time as an illusion is counterintuitive, but as we have seen in previous chapters, so are many other concepts.

Euphoria and Oneness of Consciousness

Sometimes euphoria and oneness of consciousness happen unexpectedly. Dr. Jill Bolte Taylor's stroke is one such example. On December 10, 1996, the thirty-seven-year-old healthy and vibrant neuroscientist had a stroke in the left hemisphere of her brain. Within four hours she lost her ability to talk, walk, read, and write. As blood oozed out from a vein near the left hemisphere, she became disoriented, scared, and confused, yet in this helpless and dangerous situation she somehow mustered enough presence of mind to seek help. With luck and sheer determination, she contacted her colleague at the Harvard Brain Tissue Research Center on the phone. But when she tried to speak, as she explains, she sounded like a golden retriever.

On the other end of the line, her colleague's words sounded to her like incoherent mumblings. Luckily, he realized she needed help, so he dropped everything he was doing, jumped in his car, picked her up at her house, and sped her to the hospital. Two and a half weeks after the stroke, she underwent major surgery to remove a golf ball–sized blood clot that had obstructed her brain's ability to function normally.

Eight years after the surgery, with the love, support, and guidance of her mother, she had recovered all her physical and mental functions. In fact, above and beyond recovery, she learned a great deal from the experience. Her stroke not only revealed to her some functions of the left and right hemispheres she hadn't known about as a neuroscientist, but led her to gain insights about who we are and the nature of reality. She discovered that the intuitive right brain, when unable to correspond with the left hemisphere, creates feelings of supreme peace and euphoria.[29] And in fact, the experience of euphoria and accompanying feelings of being in oneness with consciousness were, according to Taylor, the most valuable insights she gained.

Since the normal left and right hemispheres of the brain function similarly for everyone, she thought her story was worth telling. In the resulting book, *My Stroke of Insight: A Brain Scientist's Personal Journey*, she describes her remarkable experience with clarity and sensitivity. Since her recovery she has given numerous talks all over the United States, and her 2008 TED talk is well worth watching. In fact, that rendition of her remarkable tale is one of the most compelling stories I have ever heard.

Before we go further, let's review the fundamentals of neuroscience and the functions of the left and right hemispheres of the brain. This will help illustrate the nature of who we are and why we function the way we do. Taylor explains that the two hemispheres of the brain have distinctly different functions. She writes, "The two hemispheres communicate with one another through the highway for information transfer, the corpus callosum. Although each hemisphere is unique in the specific types of information it processes, when the two hemispheres are connected to one another, they work together to generate a single seamless perception of the world."[30]

Among other faculties there are specific areas in the brain for movement, speech, vision, hearing, smell, taste, and physical boundaries. When one of these areas is damaged, it causes problems specific to that site. But the presence of specific sites in the brain doesn't mean that these areas *create* speech, vision, or hearing. These locations in the brain receive information from the external world at the energy level of vibrating and spinning atomic particles. We are literally enveloped in a sea of electromagnetic fields, and our sensory organs in conjunction with the brain create what we experience. In other words, the sense organs and the brain create images of the objects we see or the sounds we hear.[31]

Since our ability to read, write, talk, and walk reside in the left brain, the damage caused by Taylor's hemorrhage left her as an infant in a woman's body. But through her persistent, sustained efforts to recover, she gained back the memories and abilities she'd had before her stroke. Now here is a question: Where had her abilities to read, write, walk, and talk and her knowledge of neuroscience gone when the left portion of her brain was damaged? Did they automatically get stored somewhere else? How did they return? What could have been the reason or purpose for Taylor to experience euphoria, ecstasy, peace, and nirvana when her left brain wasn't functioning?

Taylor explains that because of her stroke, she gained perspective about who we are and about consciousness that she hadn't had prior to her stroke. She suggests that her "stroke of insight" allowed her to reinterpret the very nature of reality and her purpose in life. She realized that she was part of a greater whole that was the source of everything in the universe. She understood that the source of deep love, peace, and serenity lies within us and that the right brain is the mechanism through which all of us experience it. Ultimately, in spite of the debilitating difficulties she encountered as a result of her stroke, she was grateful for it.

Why would a neuroscientist from Harvard come to such conclusions? Why would she risk her credibility by telling the world about her new insights and perspectives, which would undoubtedly be dismissed or even ridiculed by many?

When people have an extraordinary experience, it is usually difficult for them to dismiss it. We have discussed neurosurgeon Eben Alexander's NDE, astronaut Edgar Mitchell's *samadhi* experience, and the ESP abilities of many others. Such experiences are transformative for those who have them. Taylor's euphoric state in the absence of the chatter of her left brain unblocked her connection to consciousness and permitted her to gain the insights she was meant to have. Her remarkable recovery affirmed that the left brain is a mechanism for decoding these messages, which originate in consciousness.

Two Domains of Consciousness

Human beings function in two separate domains. I will refer to them as the physical domain, or Self-1, and the infinite domain, Self-2.

Self-1 experiences everyday reality; it requires food, water, shelter, rest, and sleep. It has emotional needs for affection, social interaction, and love. It experiences pain and pleasure. It is incomplete and ego driven.

Self-2 is perfection. It requires nothing. It is pure love and wisdom. It is the source of our conscience. It *is* consciousness.

Maintaining balance between Self-1 and Self-2 is key to a productive, meaningful, and purposeful life. Yet most of us often fail to maintain balance between these two selves.

For example, while I was writing this book I had to balance my time between work and a healthy lifestyle. Whenever I failed to get enough sleep, did not eat properly, exercised insufficiently, or had minimal social interaction, I was unhappy and unproductive.

Though the need to live a balanced life had been obvious to me for a long time, the new awareness of being a composite of Self-1 and Self-2 made it easier for me to regain and maintain balance in life. Instead of being critical of my inability to work longer or be more productive, I recognized the importance of attending to the needs of my Self 1. I became more compassionate toward myself (Self-1), and willingly took the time to rest, sleep, spend time in nature, meditate, and exercise sufficiently.

Self-2, which exists within all of us, is perfection, pure love, and wisdom. It is why we are supreme beings capable of infinite knowing, doing, experiencing, and creating. This kind of thinking may be met with disbelief, considered irrational and outside the realm of science. It may also be ridiculed or ignored by many—but not by everyone, for there is considerable scientific evidence of our supreme capacities. You will recall that in chapter 3 we saw the extraordinary musical capacity of a twelve-year-old, the inexplicable abilities of savants, and extrasensory perception among ordinary people.

Our dominant perceptions and day-to-day functioning are in the realm of our physical self, though the infinite self is primary, remains in the background, and guides the entire show of our existence. Access to the power of infinite self is not readily available to Self-1, but in meditation or prayer it's often possible to experience its wonders. Sometimes this infinite power seems to come into our awareness without any conscious effort on our part. For example, my ability to read a stranger's mind and see in the dark when I was a freshman in college happened spontaneously. Even decades later, it doesn't seem to be a random event without purpose.

In the next chapter we will explore our infinite capacities and learn how to consciously use them to live more purposefully and with greater ease. We were meant to free ourselves from the caves of our psychic prisons and discover the world with new insights. In our discovery we may begin to live more in accordance with how things are and were meant to be. Greater purpose and a deeper sense of connection with everyone and everything in the universe are possible when we align ourselves with the true nature of reality.

Now let's review the ways in which the system of logic we use is an impediment to understanding the nature of reality and consciousness.

The Tetralemma Logic of Nagarjuna

Many have found the astonishing findings of quantum theory regarding the nature of reality difficult to understand because human thought and language, and the logic they are founded upon, are *binary* in nature. This binary system of logic has evolved in the framework of physical reality. Everything we see, hear, feel, and know is based on the three-dimensional reality of the physical world and our five senses, upon which human thought processes and language are based. Therefore, phenomena in nonlocality (beyond space and time) and the subatomic world are beyond what the binary system of logic can explain. However, amazingly, a more comprehensive system of logic exists: tetralemma logic.

According to the interpretations of the scriptures of Buddhism, the theory of *emptiness* is attributed to the historical Buddha. But according to the Dalai Lama, "This theory of emptiness was first systematically expounded by the great Buddhist philosopher Nagarjuna (c. second century CE). Little is known of his personal life, but he came from Southern India and he was—after the Buddha himself—the single most important figure for the formulation of Buddhism in India."[32]

Nagarjuna's explanation of "emptiness" requires a more expansive way of thinking, analyzing, reasoning, and interpreting the nature of reality, which he explicated in his tetralemma logic. In contrast to the binary or Aristotelian logic of true and false, the tetralemma logic of Nagarjuna broadens our thinking and allows us to "conceptualize" nonconceptual phenomena. It explains that statements about the nature of reality can be:

1. True
2. Not true
3. True *and* not true
4. Neither true *nor* not true (Nagarjuna believed this was usually the case)[33]

The first two parts, true and not true, are the same as in binary logic. For example, it is true that at this moment you are reading, and it is not true that you are scuba diving.

The third and fourth parts of this system of logic are *nonconceptual*. For example, the concept of infinity is understandable but nonconceptual. Similarly, though it isn't possible to conceptualize *true* and *not true* as existing simultaneously, it is possible to get a sense of their nonconceptual nature. For example, the nonconceptual statements "You are alive and not alive both at the same time" or "You are in your house and on the moon at the same time" give us a sense of this nonconceptual statement of true and not true simultaneously.

To help us gain a sense of the fourth part of the tetralemma logic, "neither true nor not true," an example would be "You exist everywhere, without existing anywhere, unless you are observed." These examples show that it is impossible to understand tetralemma logic in the framework of binary logic.

This system of logic is "non-conceptual and a nondual path between reductionist materiality and nihilism. Materiality and nihilism can be seen as Aristotelian logical views of a single truth. The materialists say that the person is simply the sum of its parts, while the nihilists say the person is an illusion."[34] Nagarjuna's logic is an open-ended philosophy that overcomes the shortcomings of both Aristotelian dualism and nihilism. Thus, the implications of tetralemma logic are profoundly sophisticated. It helps us understand the limitations and inaccuracies that are embedded in our language, thought patterns, and conceptions of the world.

Besides a philosophical perspective, is there any evidence of the accuracy of Nagarjuna's four-part system of logic?

Interestingly, scientific evidence for it does exist, discovered in the 1930s with the double-slit experiment of quantum theory we discussed earlier. This experiment demonstrated that particles appear to be waves, particles, both waves and particles at the same time, and neither waves nor particles, depending on the observer. Most scientists for almost the last hundred years have tried to understand these observations within the limitations of binary logic, and are therefore baffled by their findings and implications. However, the behavior of subatomic particles *mirrors* the four-part logic of Nagarjuna.

An animated version of the way particles and waves behave can be found on YouTube under the title *Dr. Quantum—Wave Particle Duality*. The fact that the behavior of subatomic particles can be understood in the context of tetralemma logic is profound, for it not only validates the accuracy of this system of logic, but it also reveals the "magical" truth about the wave-particle duality determined in the double-slit experiment.

Since the universe is made of subatomic particles, by extension it is reasonable to assume that the universe also behaves like the particles that make it up. And, when we review it in accordance with tetralemma logic, we can conclude that it has a mind of its own. We refer to this mind of the universe as *consciousness*.

In the twentieth and twenty-first centuries, scientists and philosophers like David Bohm, Deepak Chopra, Amit Goswami, Ervin Laszlo, Dean Radin, and others have contributed significantly toward a better understanding of the nature of reality by regarding consciousness as fundamental to all phenomena in the cosmos. Today, biologists, chemists, mathematicians, medical doctors, neuroscientists, psychologists, philosophers, and physicists from all over the world are creating shifts in their respective fields by incorporating the subjective and experiential realities of consciousness into their thinking and research, and finding that reality isn't what we thought it was.

Human beings come with the gifts of intuitive intelligence, the power of the subconscious mind, and a sense of wonder, curiosity, and imagination, among other qualities. Therefore, it is possible for those of us who are not quantum physicists to understand the salient features of the quantum world, because in essence we have known that world all along. All cultures throughout history have produced stories, myths, rituals, ceremonies, and songs that are testimonials to our deep, innate inklings about our magical universe, oneness, and the nature of reality.

Sensing Consciousness

Knowledge is not merely an intellectual phenomenon. It is also intuitive and experiential—we feel the rightness of this kind of knowing in the depth of our hearts. It is the joy of witnessing a hummingbird come out of nowhere and flutter among the flowers. The unexpected delight and intrigue we feel in the presence of the tiny, colorful bird, which blends in with the colors and plumage of the flowers, leaves us in awe and wonder. This is not an intellectual knowing, but an innate and fundamental part of who we are. We simply know this bond. World-renowned biologist E. O. Wilson called this deep bond with all of nature *biophilia*. However, it is not limited to the physicality of nature. This bond is *who we are*, and it is the pervasive, singular, and ultimate nature of reality.

It's the feeling we experience at the shore of a lake, river, or ocean. The sound of waves lapping on rocks and pebbles is soothing, and we want to remain there. The sight of blue skies and the majestic beauty of the water along a tree-lined shore is peaceful, and somehow we connect with all that we see, hear, and feel. Without any thought or conscious effort on our part, we experience a union with everything around us. We may realize that this knowing and the peace that comes with it reflect our true nature. When we contemplate who we really are, we may begin to realize that we are supreme beings in union with a miraculous universe, as Deepak Chopra and physicist Menas Kafatos described in their book, *You Are the Universe: Discovering Your Cosmic Self and Why it Matters*. We may begin to understand that this is what we have been searching for. Scientific evidence of this union has helped me understand my bonds with the universe intellectually, but my intuitive bonds have always tugged at my heartstrings.

One of my favorite songs, which expresses this innate wonder and the desire to experience oneness with the universe, is expressed in "The Rainbow Connection" by songwriters Kenny Ascher and Paul Hamilton Williams. If you listen and sing along, as I sometimes do, you, too, will experience a sense of longing, joy, and wonder about our cosmic connection. When my children, Daniel and Michael, were young, I loved hearing this song with them. The memories of snuggling up with them and watching Kermit the frog sing remain precious. Back then, I heard the lyrics and wondered what the boys might be thinking and feeling. Today when I hear this song, I am moved by my old memories as well as a sense of mystical connection with our magical universe.

Let's see what it means to discover the rainbow connection—or the nature of reality—not only through science, but also intuitively, experientially, and spiritually.

Encounter with Consciousness

"Consciousness" is notoriously difficult to define because it is so fundamental; it is the precondition of our being as well as that of all phenomena of the infinite nature of the cosmos, known and unknown. So instead of trying to define the undefinable, we will focus on knowable and experiential realities that can be interpreted by our senses. Consciousness sometimes makes us aware of our oneness with it when we least expect it. When yogis, gurus, seasoned practitioners of meditation, ordinary people, or scientists experience the infinite love, unity, and expansiveness of consciousness, they are forever transformed, and they describe their accounts of their transformative experiences (referred to as *samadhi* in Sanskrit) in strikingly similar ways.

In quantum theory, the concept of *entanglement* among the subatomic particles that make up the universe is similar to the idea of oneness as described by the ancient wisdom traditions. To get an intuitive and experiential understanding of reality, life, and the universe, it is useful to review oneness—or entanglement—as astronaut Edger Mitchell, who walked on the surface of the moon, experienced it.

On his journey home, while silently observing the heavens from the space capsule's window, he had a transcendent experience in which he felt one with the entire universe. And he came to know reality and the universe differently from what he had studied in science and believed in all his life. Passages in his book, *The Way of The Explorer: An Apollo Astronaut's Journey Through the Material and Mystical Worlds*, describe his experience on his return to Earth as his two colleagues, Stu Roosa and Alan Shepard, slept.

A wonderful quietness drifted into the cabin, the satisfying glow of the job well done. The lion's share of my work was complete, and all I had to do was monitor the spacecraft systems, which were functioning perfectly. Now there was time to quietly contemplate the journey. I could lie back in weightlessness and watch the slow progress of the heavens through the module window. My mind ebbed into that quiet state I had longed for on our trek to the rim of Cone Crater. There was tranquility, and a growing sense of wonder as I looked out the window, but not a hint of what was about to happen.

Perhaps it was the disorienting, or reorienting, effect of our rotating environment, while the heavens and earth tumbled alternately in and out of view in the small capsule window. Perhaps it was the air of safety and sanctuary after a two-day foray into an unforgiving environment. But I don't think so. The sensation was altogether foreign. Somehow I felt tuned in to something much larger than myself, something much larger than the planet in the window. Something incomprehensibly big ... Even today, the journey still baffles me.... But the tableau is so vivid as to have lost none of its clarity. It looms in my memory with extraordinary resolution.... There was a startling recognition that the nature of the universe was not as I had been taught. My understanding of the separate distinctness and the relative independence of movement of those cosmic bodies was shattered. There was an upwelling of fresh insight coupled with the feeling of ubiquitous harmony—a sense of interconnectedness with the celestial bodies surrounding our spacecraft. Particular scientific facts about stellar evolution took on new significance.

Our presence here, outside the domain of the home planet, was not rooted in an accident of nature or in the capricious political whim of technological civilization. It was rather an extension of the same universal process that evolved our molecules. And what I felt was an extraordinary personal connectedness with it. I experienced what has been described as an ecstasy of unity. I not only saw the connectedness, I felt it and experienced it sentiently. I was overwhelmed with the sensation of physically and mentally extending out into the cosmos. The restraints and boundaries of flesh and bone fell away.[35]

This description is strikingly similar to words spoken by spiritual men and women who have experienced *samadhi* and the spaciousness of oneness. The experience transformed Mitchell, and he devoted the rest of his life to finding scientific answers to the nature of reality. He went on to found the Institute of Noetic Sciences, where scientific research has produced new understandings about reality that are in alignment with the descriptions found in spirituality.

The epiphany Mitchell had in space was stunning, but equally remarkable was his quest of more than two decades to investigate and understand what he had experienced. One insight he gained from his transcendent experience is worth repeating: "*There was a startling recognition that the nature of the universe was not as I had been taught.*" The fact that a navy pilot, hardnosed astrophysicist, and seasoned astronaut expressed this view grabbed my attention.

Mitchell gave up his career at NASA to devote his time and energy to the growth and development of his new institute, and to writing *The Way of the Explorer*, in which he developed a framework for understanding life and the universe based on mysticism and science. In the book, he offers convincing explanations for nonordinary phenomena such as telekinesis, telepathy, and clairvoyance, some of which we discussed in chapter 4. His findings shed light on the incredible intelligence and extraordinary capacities of human beings. Though we may be hesitant to accept our superhuman abilities as normal, nonetheless, at a deep, subconscious level we know that we are more than just temporary biological beings. And that our lives have greater purpose and meaning than we may consciously realize.

Earlier in his career, Mitchell may not have fully understood all the reasons why he had chosen a life trajectory that sent him to the moon. But after experiencing himself as one with the universe, he knew he was meant to gain knowledge of the external world so he might eventually journey into the inner world—and tell the rest of us about it.

The Nature of Consciousness

In chapter 1 we learned that neurosurgeon Eben Alexander's near-death experience led him to an awareness of a deeper and more profound nature of consciousness than he had previously understood. And as happened with Mitchell, the experience inspired him to dig deeper. So he went on to form the Eternea organization (Eternea.org). On its website he posted seven core statements:

1. The enduring essence of consciousness extends beyond the brain, transcending it and capable of existing independently of it;
2. This aspect of consciousness is eternal in nature, unbounded by space, time, and matter, and

is able to manifest in other forms and places throughout the spectrum of eternal existence;

3. All things in the cosmos are interconnected at the quantum level, influencing each other nonlocally and instantaneously, implying that all things are one in the grand web of creation;

4. The meaning and purpose of all existence and the organizing principle of the cosmos which drives the evolution of all things is to become greater expressions of harmony and love … loving all things unconditionally, including oneself;

5. There is a profound Intelligence or Source underlying the creation and evolution of the universe from which all things originate and to which all things return;

6. In an interconnected universe an intricate matrix of cause-effect relationships exists, suggesting that what we do to others we do to ourselves, which means that we reap what we sow; and

7. The good of the one and the good of the whole are mutually enhancing, affirming the ancient wisdom that the quality of both individual and collective existence is enhanced by bringing every aspect of creation into a state of complete unity, harmony and love.

There are many scientists, philosophers, and spiritualists whose work and understandings about consciousness provide deeper insights into the nature of reality and the cosmic intelligence behind it all. As I described earlier, my curiosity about the true nature of reality was triggered by my near-death and extrasensory perception experiences. Today I consider them normal human experiences when seen in light of consciousness.

The evidence of consciousness being the fundamental nature of reality is so substantial that in February 2015, eight world-renowned scientists, doctors, and researchers got together at Canyon Ranch in Tucson, Arizona, and wrote "Manifesto for a Post-Materialist Science," which was first published in *Explore: The Journal of Science and Healing*. The points made in the manifesto are as follows:

- Mind represents an aspect of reality as primordial as the physical world. Mind is fundamental in the universe; i.e., it cannot be derived from matter and reduced to anything more basic.
- There is a deep interconnectedness between mind and the physical world.
- Mind (will/intention) can influence the state of the physical world, and operate(s) nonlocally, i.e., it is not confined to specific points in space, such as brains and bodies, nor to specific points in time, such as the present.
- The shift from materialist science to post-materialist science may be of vital importance to the evolution of the human civilization. It may be even more pivotal than the transition from geocentrism to heliocentrism.[36]

Seeds of Curiosity

Like most children, when I was a kid I loved hearing stories. One day, when I was about seven or eight years old, a family friend whom I referred to as my uncle said, "Nuri, would you like to hear a story?"

"Yes, indeed!" I answered with enthusiasm and delight.

"Great," he said. "I'll tell you the best story I know."

Now I was even more excited to hear this story.

"In the beginning," he began, "there was no earth, moon, sun or stars. There was simply nothingness. Then out of nothingness, just by chance, some fourteen billion years ago there was a huge explosion, which scientists call the Big Bang.

"A big bang?" I asked excitedly.

My "uncle" Mustafa Maamu smiled at my enthusiasm and went on to tell me how the solar system came to be, and that in the beginning the earth was red-hot molten lava, and that after hundreds of thousands of years it started to cool—and the rest of the story, with which you are familiar.

Mustafa Maamu was a second-year medical student at the University of Dhaka in Bangladesh, and he sounded well informed as he spun the tale.

He told me about the single-cell amoeba born in the rich nutrients of the oceans, which were formed by continuous rainfall upon the young planet. From the single cells, more complex life evolved. I was thrilled to hear that some fish developed arms instead of fins, and ventured on land, and later that dinosaurs roamed the earth.

Well, by the time he had finished telling me the Big Bang theory and the theory of evolution, I could hardly contain my excitement. I thanked my uncle and said, "Mustafa Maamu, this is the best story anyone has ever told me!"

Reality Reexamined

Often, certain understandings assume indisputably factual status. But then, under the scrutiny of new evidence, these previously held beliefs and dogmas are revised and clarified.

We crave the comfort of certainty in an uncertain world, but this certainty comes at a price. Deterministic understandings about life and the universe are in many ways at the root of our existential angst, fear, and discontent. Embracing new scientific findings about the nature of reality with humility and wonder, appreciating their magical qualities, and recognizing their uncertainty and ambiguity grant us freedom to reach deeper wisdom, peace, and love.

By this I do not mean we should all go to "la-la land" to attain peace and harmony. Curiosity and a desire to know more are natural for us, so we should of course continue to strive for better understandings. But not without paying attention to ancient Eastern philosophies and new findings in science, listening to the silent longings of the soul, and relying more on intuitive intelligence and the infinite knowing of our universal mind.

I invite you to continue to engage with this chapter and the rest of the book in a spirit of curiosity and openness, and with a sense that the universe reveals its wisdom and love when we set aside our preconceived notions and beliefs. As I mentioned earlier, I am not trying to prove any scientific or spiritual perspectives; rather, my intention is to reach your curiosity and your sense of wonder so you may arrive at your own deeper understandings. This more profound type of knowing already exists in you. You are the infinite mind, or the power within! This doesn't require proof but rather a recognition and remembrance of your magnificence, love, and wisdom, for that is who you are and have always been.

The story my uncle told me many years ago was the science we knew at the time, and it is still what most mainstream scientists believe today: they are certain that the birth of the universe happened by chance. However, today the story needs to be updated to reflect new findings. And the new and improved version of this story is far more compelling. We will look at it next, as well as at a few other previous scientific "facts": not to disprove them, but to become aware of their shortcomings so we may think differently and reach for deeper truths.

Neither Unique nor by Chance

Mainstream scientists suggest that evolution occurred on our planet through a series of fortunate coincidences. They are right about the "Goldilocks" conditions of Earth's location in the solar system. It is neither too far from nor too close to the sun; it has the right amount of water for producing and sustaining life; the water is both above and below its surface; the temperature variation on the surface is within the range required for life; it has the right oxygen-nitrogen rich atmosphere; and its nearly circular orbit with a twenty-four-hour rotation is suitable for life. Earth's magnetosphere protects life from radioactivity and dangerous solar radiation. Its position in proximity to huge gas planets provides protection from asteroids. Indeed, these, among others, are perfect conditions for life on Earth to evolve. However, this is not a unique phenomenon in the Milky Way. Recently, scientists have identified more than two thousand planets of similar size around suns of similar intensity in our galaxy.[37]

In his book *The Intelligence of the Cosmos: Why Are We Here? New Answers from the Frontiers of Science,* Ervin Laszlo writes that scientists working with the Kepler space telescope have found:

> On the average, each star in the Milky Way galaxy has at least one planet, and one in five 'sun-like' stars is likely to have an Earth-size planet in the Goldilocks zone. With 200 billion stars in our Milky Way galaxy, there could be 11 billion Earth-sized planets in the Goldilocks zone in this galaxy alone—and there are 10^{22} to 10^{24} galaxies in the universe—this is highly probable because the evolution of coherent systems (not random, but highly ordered) is not due solely to fortunate conditions.[38]

Laszlo explains that there is deep intelligence in the universe. To substantiate his assertion, he adds that the universe is amazingly precise, coherent, and mathematical in nature. Richard Feynman, a Nobel Prize winner in quantum electrodynamics, conveyed a similar understanding when he said, "Why nature is mathematical is a mystery ... The fact that there are rules at all is a kind of miracle."[39]

To explain the unlikelihood of the universe coming into existence by chance, astronomer Sir Fred Hoyle, after examining the intricate, precise, and mathematical accuracies of various systems such as gravitational force, radioactivity, electromagnetism, radio waves, and microwaves, among other parameters of the universe, explained that the probability of the universe to have emerged out of nowhere by chance is equivalent to a tornado passing through a junkyard leaving behind a fully functioning modern-day jetliner.[40]

In the middle of the twentieth century, astrophysicist Sir Arthur Eddington and theoretical physicist Paul Dirac, who was awarded the Nobel Prize in physics in 1933, also noted curious "coincidences" concerning the parameters of the universe. Are there new findings in science that support Hoyle's metaphorical explanation? And what were some of the specific scientific "coincidences" Eddington and Dirac noted that cast doubt on the idea that the universe emerged by chance? Let us examine a few of them.

The ratio of electrical force to gravitational force is 10^{40}, and, interestingly, it is nearly the same as the ratio between the size of the universe and the size of subatomic particles. This is surprising because the electrical and gravitational forces are constants (unchanging), whereas the ratio between the size of the universe (which is expanding) and elementary particles is not, and therefore it should continuously change. Yet that is not the case. This made Dirac speculate that because the ratio of 10^{40} remains nearly the same when it should change, it is more than random coincidence. This suggests that either the universe is not expanding (which it is) or that the force of gravity varies proportionally to its expansion. Laszlo asserts that facts like these suggest the presence of intelligence in the universe.[41] Interestingly, there are many more precise and self-regulating systems in the universe that point in the same direction.

For instance, another astonishing coincidence that suggests that the universe is mathematically and precisely designed can be deduced from the examination of harmonic ratios of the mass and number of elementary particles and the forces between them. Even microwave background radiation, which is a remnant of the Big Bang, is unexpectedly coherent. According to Laszlo, "It is dominated by a large peak followed by smaller harmonic peaks." He goes on to explain that these "series of peaks end at the longest wavelength, which physicist Lee Smolin termed R."[42] Interestingly, when R is divided by the speed of light (R/c), it gives us the age of the universe. And when R is squared and divided by the speed of light (R^2/c), it gives us the rate of expansion of the galaxies.[43]

These are astonishing "coincidences"! They tell us that the parameters of the universe are finely tuned. More than two dozen other remarkable mathematical relationships exist in the universe, making it absurd to conclude that the universe happened by chance and is without purpose, meaning, or design. But when it comes to blind faith, no matter what the empirical evidence may suggest, believers of scientific materialism do not change their minds. Geoffrey Madan, the English writer, once said, "The dust of exploded beliefs may make a fine sunset."[44]

"Big Bang" Reexamined

In their book *CosMos: A Co-creator's Guide to the Whole World,* Ervin Laszlo and Jude Currivan reexamine the orthodox understandings of the Big Bang theory and point out several significant loopholes and assumptions about it. Reviewing recent findings in astronomy and astrophysics, they point out that mainstream contemporary explanations of the origins of the universe, and the fundamental nature of reality, are not entirely accurate.

Laszlo and Currivan explain that 96 percent of the universe is "missing," and point to several discoveries that shed new light on the origin and nature of the universe. In the early 1930s, Edwin Hubble, after whom the Hubble space telescope is named, "analyzed light from 24 distant galaxies and discovered that the galaxies were not only moving away from our own, but the further they were from us, the faster they were moving away." Laszlo and Currivan explain further: "Realizing that this meant the universe itself was expanding, other astronomers drew the logical implication that by going back in time, the universe would shrink to a primeval origin of enormous temperature and density."[45]

Thus the birth of the universe as a "Big Bang" was

introduced for discussion and debate and was eventually to become the dominant view of astronomers, scientists, and the public at large. And scientists have found evidence to support the Big Bang model. For example, Laszlo and Currivan explain:

> The observed microwave energy field that pervades the universe is deemed to be the residual energy signature of its birth. And the [Big Bang] theory has been used to model how the subatomic particles that make up the material world were derived in the first few minutes after the Big Bang.
>
> But the model also requires a period of almost unimaginable rapid inflation of space moments after the bang to explain the regular distribution of galaxies observed across vast tracts of the universe. Without such inflation, the model is incapable of showing how such extended contact could have occurred—inflation whose cause, mechanism, and reason for stopping as abruptly as it began also remains unknown.[46]

Even as concerns about the validity of the Big Bang emerged, until 1998 the Big Bang theory, modified to include inflation, remained undebatable. A basic premise of the theory is that the rate of expansion of space will eventually slow down as the force of gravity gradually overcomes the force of the initial bang. However, in 1998 this assumption was discovered to be wrong. Two independent teams of astronomers observed that the expansion of the universe seems to be accelerating. This is metaphorically equivalent to observing that a rock thrown up in the air does not fall back to the ground, but rather, accelerates upward the higher it goes. Interestingly, the energy associated with the acceleration remains a mystery and has not been observed directly. Furthermore, the assumed *dark energy*, which should be approximately three-quarters of all matter and energy of the universe, does not come anywhere near the measured value. It's off by a factor of 10^{200}! According to Laszlo and Currivan, this discrepancy has been called "the greatest embarrassment in the history of science."[47]

Furthermore, to explain the stability of the galaxies, the concept of *dark matter* had to be introduced. Of course, this is also an assumption. However, without this assumption we cannot claim the Big Bang theory is accurate. Even after introducing the assumption of dark matter, we can only explain 4 percent of the matter and energy in the universe. The other 96 percent of dark matter and dark energy are simply missing and we have no explanations for why that would be.[48]

So, what are the implications of this? Simply put, the Big Bang theory is not a definitive answer to how the universe began, nor do we know how it works. This realization among physicists, mathematicians, and cosmologists has spawned many alternative theories about the universe. Brian Greene, professor of physics and mathematics at Columbia University, in his book *The Hidden Reality: Parallel Universes and the Deep Laws of the Cosmos*, writes, "What we've found has already required sweeping changes to our picture of the cosmos," and provides details of alternative theories such as multiverses, parallel universes, and holographic universes.[49]

I believe the driving force behind contemporary alternative theories about the universe is the human desire to have concrete and definitive answers. But it is impossible to arrive at the answers to how the universe started, where it's headed, and what will happen to it in the distant future. Concepts of beginnings and endings are rooted in the binary logic of all languages and the manner in which we think and interpret reality. Therefore, contemporary observations, calculations, assumptions, and conclusions about the Big Bang theory, as well as many other contemporary alternative theories about the universe and the nature of reality, shall remain mysterious in the context of ordinary (binary logic) thinking. But, as we discussed earlier, we will not remain stuck forever in seeing reality from this limited perspective. When we review the nature of reality in the context of the ancient wisdom traditions, tetralemma logic, and the expansive understandings of quantum theory, we will have greater awareness about the universe and who we really are.

In addition to using an incomplete system of logic, the problem with the Big Bang theory is that it only takes into account energy and matter. Because consciousness, which is the bedrock of reality, is not taken into consideration, the elaborate mathematical attempts to explain the birth of the universe end up providing confusing, inaccurate, and sometimes outlandish explanations. However, when consciousness is seen as the source of all phenomena, we can arrive at better understandings.

Did Life Originate in the Stars?

Let's revisit the story my uncle told me when I was seven years old. In this fascinating tale I learned that life originated in the oceans, which were created by heavy rains lasting for thousands of years. I also learned that the building blocks of life—carbon, hydrogen, and oxygen—existed on Earth, and that the combining of these elements into different organic compounds somehow gave birth to life. The story my uncle told me contained truths, along with a major assumption that life must have originated on Earth. This assumption has led scientists to attempt to create life in the laboratory, but to no avail.

In 2010 NASA scientists found evidence of sugar molecules and other organic compounds in the white-hot crucible of stars in the distant galaxies of Andromeda and Triangulum. The most widespread group of molecules found were the polycyclic aromatic hydrocarbons (PAHs). This is amazing, because these galaxies are very different from our Milky Way.

These organic compounds remain stable in spite of the enormous heat of the stars, and withstand high levels of radioactivity, thousands of times higher than what we are exposed to on earth. These organic materials exist not only in the stars and planets, but also on asteroids, which travel through space and pollinate the entire cosmos.[50]

The existence of these building blocks of life in distant

galaxies does not in and of itself disprove that life originated on planet Earth; rather, it raises interesting questions and speculation about the origin of life.

For example, life may well have originated in the stars. After all, iron, gold, copper, and other elements in our bodies are the same elements that existed in the stars long before we came along. You would not be alone if this new information from NASA about the existence of sugar molecules in distant galaxies generates a greater sense of wonder and curiosity about the nature of our cosmos, who we are, and how we got here.

Over the last decade, in addition to NASA's findings, many scientists throughout the world have gained further evidence of the existence in distant stars of organic molecules that are the building blocks of life, including the existence and spontaneous emergence of DNA. Laszlo notes, "A team of astrophysicists headed by Sun Kwok and Yong Zhang at the University of Hong Kong found 130 macromolecules present in the vicinity of active stars. They included glycine, an amino acid, and ethylene glycol, the compound associated with the formation of the sugar molecules necessary for life. Their presence suggests that they were ejected in the course of the stars' thermal and chemical evolution."[51]

Furthermore, organic molecules were also discovered in interstellar clouds. Arnaud Belloche and his colleagues at the Max Planck Institute for Radio Astronomy in Bonn, Germany, published a paper in October 2014 in which they documented the presence of *iso*-propyl cyanide (i-C_3H_7CN). The branching carbon structure of this complex molecule is similar to the amino acids that form the basis of proteins on Earth.[52]

The most astonishing evidence of consciousness from which the universe emerged was documented in a study published in *Science* by Tommaso Bellini of the University of Milan and Noel Clark of the University of Colorado, Boulder. This study documented the spontaneous self-assembly of DNA fragments, only a few nanometers in length, into ordered liquid crystal phases, which have the ability to drive the formation of chemical bonds that connect together short DNA chains to form long ones, without the aid of biological mechanisms. Along with the spontaneous formation of DNA, the study also documented the non-biologic origins of nucleic acids, the building blocks of living organisms.[53]

Stephen Hawking said, long before the discovery of organic compounds in distant galaxies and the discovery of the spontaneous self-assembly of DNA, that simply because there are so many galaxies, stars, and planets in this universe, it is inconceivable that life would not be abundant everywhere in the universe. It is evident to many scientists and others that the universe is teeming with life and that everything everywhere is a representation of consciousness.

Let us for the moment leave aside the origin of life, and instead dig deeper to see how cells grew into more complex organisms and our own biology. Taking this deeper dive will help us further understand that the prevailing understandings rooted in scientific determinism have led us to many wrong conclusions.

Does DNA Determine Destiny?

The search for certainty in an uncertain world has engendered various strains of determinism throughout history. As our understanding of genetics and genomics has grown, genetic determinism has taken hold among many physicians, scientists, and the general public, who believe that the genes we inherit from our forebears become our biological destinies.

Our inherited genes and the changes that occur in them can affect our lives dramatically. They may predispose us to various degenerative and metabolic diseases, or toward cancer, and when they do they create profound challenges and choices for us. However, new understandings about genes gleaned from the Human Genome Project, the newly discovered functions of cell membranes, and the field of epigenetics suggest that genetic predispositions toward disease are not the sole determinants of our biological destinies.

It is estimated that the human body consists of 50 trillion cells, which are made up of four types of large molecules: polysaccharides (complex sugars), lipids (fats), nucleic acids (DNA/RNA), and proteins. Though cells require each of the four molecular types, proteins are the most important components, and our bodies require more than 100,000 different types of proteins in order to function. Respiration, digestion, and muscle contraction are regulated by protein molecules, which strive to maintain balance between their positive and negative electromagnetic charges as they constantly interact with other molecules, hormones, enzymes, or the interference of electromagnetic fields such as cell phones and computer monitors. The intrinsic quality of the protein molecule to reestablish balance in its electromagnetic charge, which can occur thousands of times in a second, is at the heart of all living organisms. It is the changing nature of protein molecules that regulates life, not deoxyribonucleic acid (DNA), as orthodox science understands it. In spite of this knowledge that we now have, the central dogma of molecular biology still claims that DNA controls biological life.[54]

One of the recent massive blows to the central dogma of genetic determinism came, ironically and inadvertently, from the Human Genome Project, which was initiated in the late '80s to create a catalogue of all genes present in humans. Conventional thinking at the time was that because there are 100,000 or more different proteins along with 20,000 protein-encoding genes in human beings, when all of the genes were found they would add up to approximately 120,000. To the utter shock of practically all geneticists around the world, the entire human genome was found to consist of "only 20,000 to 25,000 genes, a surprisingly low number for our species."[55]

As surprising as this finding has been for the one-gene-for-one-protein dogma, the number of genes was only lower than expected because of the false understanding that genes determine destiny. The low number of genes in humans is even more astonishing when it is compared to the number of genes found in frequently studied animals like the mouse, the fruit fly, and the primitive nematode round worm. Interestingly, the mouse, with approximately 23,000 genes, would have 3,000 more genes than a human being when the lower estimate is used, or 2,000 fewer when the higher estimate of genes in a human is used. How can this be possible? Furthermore, the fruit fly, a more advanced animal than the nematode worm, has 15,000 genes, and the primitive nematode worm has 24,000 genes. How can a more advanced fruit fly have approximately 9,000 fewer genes than the worm? And, how can a human body comprising over 50 trillion cells contain approximately the same number of genes as the microscopic worm with 969 cells?[56]

Nobel Prize winner and preeminent geneticist David Baltimore, upon reviewing the findings of the Human Genome Project, said, "But unless the human genome contains a lot of genes that are opaque to our computers, it is clear that we do not gain our undoubted complexity over worms and plants by using more genes. Understanding what does give us our complexity—our enormous behavioral repertoire, ability to produce conscious action, remarkable physical coordination, precisely tuned alterations in response to external variations of the environments, learning, memory, need I go on?—remains a challenge for the future."[57]

Dr. Baltimore did not explicitly give any reason why the one-protein-to-one-gene dogma had collapsed, but his comment implies that scientists may have been surprised by the findings of the Human Genome Project because they had failed to consider the role of consciousness in the workings of DNA.

Based on the fact that humans have approximately 100,000 fewer genes than had been anticipated, it is reasonable to conclude that genes encoded in the DNA strand do not determine human destiny. Though some functions of the human body, such as our height or the color of our skin, seem to be hardwired, we have known intuitively and scientifically that the environment in which an organism grows influences it. In humans a mother's physical, emotional, and psychological health influences the developing fetus. After birth the development and well-being of a child are significantly affected by parental love and guidance. The interactions of family, friends, habits, and personal and cultural values all influence our health and well-being throughout life, and so do our thoughts and feelings. DNA is ever changing in response to the signals in the environment. Bruce Lipton in *The Biology of Belief* notes, "Every protein in our bodies is a physical/electromagnetic complement to something in the environment. Because we are machines made out of protein, by definition we are made in the image of the environment, that being the Universe ..."[58] Lipton's comment is another way to state that consciousness is the source of all there is. It is what makes DNA and proteins function as they do. Human thoughts, emotions, and feelings are integral parts of consciousness that constantly shape and reshape our DNA. We are indeed the co-creators of our realities.

Science and Spirituality

Why has there has been so much controversy between science and spirituality? At the heart of it are the differing approaches science and spirituality have taken in understanding reality. In spirituality, we take an inner journey to arrive at deeper understandings. In a state of quiescence, we enter into the quantum/nonlocal realm of greater knowledge. In fact, the greatest discoveries in science as well as creativity in art, music, and literature occur in this domain.

Science, on the other hand, is rooted in materialism, binary logic, and scientific inquiry based upon measurement and repeatability. Pursuit of knowledge based upon the scientific method of inquiry has produced stupendous results, including technological marvels, knowledge of the physical universe, and discoveries in physics, chemistry, and biology, to name a few. These discoveries have shaped the modern world. But mainstream scientists do not recognize the limitations of the scientific method, and this has led to many inaccurate conceptions and understandings.

The findings in quantum theory about the nature of reality uprooted basic assumptions of twentieth-century science. Albert Einstein, upon reviewing these new findings, remarked that they meant we would have to throw out everything we had come to accept in all of science. Einstein's statement signified the relevance of the material world, but the implication of his statement was an either/or proposition, for he wasn't prepared to fully accept the findings of quantum theory.

Today, like Einstein, most scientists are still stuck in materialism and binary logic. Yet a good number of them are not. In fact, many contemporaries of Einstein continued to explore quantum theory, and today substantial proof exists regarding its validity. What I have shared with you are findings of the scientists who have pursued knowledge and truth regardless of where it took them. Among mainstream scientists, there are many who set out to prove the findings and theories of the new science wrong but instead became converts, either willingly or with some reservations. Others pursuing normal scientific methods found astonishing results and reviewed their findings afresh.

Significantly, in the earlier part of the twentieth century, quantum physicists concluded to their amazement that not only was the prevailing understanding of reality as mechanistic and deterministic wrong, but the findings of quantum physics were in harmony with explanations provided by the sages of ancient India. Astrophysicist Sir Arthur Eddington came to the conclusion that our minds and matter are interconnected. His interpretations were based on both the principles of quantum theory and spiritual understandings. He said, "If I were to try to put into words the essential truth revealed in the mystic experience, it would be that our minds are not apart from the world; and the feelings we have of gladness and melancholy and our yet deeper feelings are not of ourselves alone, but are glimpses of a reality transcending the narrow limits of our particular consciousness ..."[59]

The interconnected nature of our minds and the universe, as renowned physicists of the twentieth and twenty-first centuries explain them, illustrate significant intersections between quantum theory and ancient mystical knowledge. One such physicist was David Bohm, who had been deeply influenced by Einstein as well as by mysticism. His main concern was with understanding the nature of reality in general and consciousness in particular. In his classic work, *Wholeness and the Implicate Order*, published in 1980, he treats the totality of existence as an unbroken whole.

Bohm and Einstein were colleagues who worked together at Princeton University. While the earlier contributions Bohm made to quantum physics were well received by Einstein, his later understandings and interpretations about the interconnectedness of mind and the universe were not. Even though Einstein had come to similar conclusions mathematically, he did not personally accept them as truth.

Bohm postulated that reality was made up of matter, energy, and (thought) *meaning*. In contrast, the orthodox description of reality in science was limited to matter and energy, as it is today among most mainstream scientists. While the debate among the old- and new-school scientists continues, for me it has been interesting and informative to review David Bohm's quantum theory because his work provides a compelling explanation of the nature of reality by bridging mathematics and mysticism.

In Bohm's interpretation, the universe is a self-regulated, intelligent entity in which matter, energy, wisdom, and all life forms are interconnected and engaged in an exquisite flowing dance of creation and change. I was thrilled to discover this interpretation made by one of the most accomplished and brilliant theoretical physicists of the twentieth century, one whom Einstein had referred to as his "intellectual successor" and about whom he proclaimed, "If anyone can do it, then it will be Bohm."[60]

Bohm's findings are expansive and liberating. His theory allows us to interpret the universality of intelligence in all forms of life—as well as nonordinary phenomena like telepathy, clairvoyance, precognition, remote viewing, and psychokinesis—as ordinary capacities that all human beings possess. These nonordinary qualities reside in our subconscious or the universal mind and can be culled from there and put to use.

In Sum

The universe is an enchanted and magical place. No mathematical or scientific factual description can capture the meaning, purpose, and mystery of this spiritual entity. We are an integral part of this creation. We were meant to experience it in its physical form, but we were also meant to have a sense of its infinite complexity, beauty, wisdom, and love.

To understand the universe in this manner is not for the fainthearted. Those of us who choose to experience the mystery and infinite grandeur of the universe must look past the veils of the physical world and see it in its energetic, vibrating emptiness and wholeness. We must also look past the dogmas embedded in the religions of the world and the limitations of scientific materialism. When we do, we open ourselves to experience the vastness, beauty, and magic of the universe. We realize that we are infinite and miraculous beings experiencing finite lives in biological form. Though we were meant to experience life in physical form, and our minds have created limitations, we have the capacity and knowledge to transcend these limitations and instead experience lives filled with joy, meaning, and abundance.

We are our own co-creators. In the next chapter we discuss how we can co-create the realities we wish to experience.

6

Peaceful Mind

*Discovering the intrinsic divinity at our core
is the highest achievement;
Dying without experiencing it is the greatest
loss.*
—Pandit Rajmani Tigunait

*Complete mastery over the roaming tenden-
cies of the mind is yoga.*
—The Yoga Sutras of Patanjali, *Sutra 1:2*

Self-luminosity is the mind's innate attribute.
—Pandit Rajmani Tigunait

Ask, and it shall be given you: seek, and ye
shall find; knock, and it shall be opened on to
you.
—Matthew: 7:7

Happy the man who discovers wisdom,
the man who gains discernment:
gaining her is more rewarding than silver,
more profitable than gold.
She is beyond the price of pearls,
nothing you could covet is her equal.
In her right hand is length of days;
in her left hand riches, and honor.
Her ways are delightful ways,
her paths all lead to contentment.
She is a tree of life for those who hold her fast,
those who cling to her lead happy lives.
—Proverbs 3:13–18

I n previous chapters I showed the ways in which several foundational understandings of science are either incomplete or entirely wrong. These inaccuracies are based on the notion that the universe consists only of energy and matter. Consciousness, which is the source of energy and matter, is deemed irrelevant or nonexistent. Yet when consciousness is considered to be the source of all phenomena, the "hard problem" is solved.

It becomes possible to understand that we are spiritual beings experiencing life in biological form, and to see the universe for what it is: a spiritual entity. When we understand the universe and who we are differently from how we have considered them to be, we are able to experience life with greater peace, love, and purpose.

In this chapter we will discuss the nature of reality and who we are from the perspective of our daily lives and spirituality. The insights I present here reflect several ideas we explored in previous chapters. This discussion of spirituality does not refer to any particular religion; rather, it delves into our innate wonder about the magnificence of our being, and the sense of "something greater" than us. It addresses the spiritual wisdom that the sages, yogis, and mystics of the world have revealed, known as the *perennial philosophy*. In the context of this spiritual or noetic knowledge, I will also present pathways to peace.

Specifically, we will look at certain aspects of the *Bhagavad-Gita*, Buddhism, and the *Yoga Sutras of Patanjali*. While the scope of this book doesn't permit a comprehensive treatment of these perennial philosophies, I believe you will find even a brief discussion of them helpful. Aldous Huxley, in *The Perennial Philosophy*, states, "Reality is not clearly and immediately apprehended, except by those who have made themselves loving, pure in heart and poor in spirit."[1] "Poor in spirit" in this case means humble in our understanding of our spiritual nature. Pride can take all kinds of forms, but the worst is spiritual pride. The more we think or believe that our existence is by chance, random and without purpose, the poorer we are in our hearts. Our innate desire to live a happier, more meaningful, and more joyous life is possible when we live in accordance with the perennial philosophy.

The Perennial Philosophy

The view held by eminent theologians, mystics, and Eastern philosophers at various times is referred to as the perennial philosophy.[2]

It is perennial because it holds profound, everlasting insights into life and nature, spanning cultures, and has been taught by the great thinkers of all time. Five important claims about reality and human nature lie at the heart of this philosophy, and they summarize the salient points of this book:

1. There are two realms of reality. The physical or phenomenal world is not the only reality. Another, nonphysical realm exists, and the two domains together constitute the ultimate reality.
2. Human beings are a composite of both these realms and mirror the nature of this two-sided reality.
3. Humans possess the capacity to perceive non-physical reality.
4. We can recognize our divine spark and know that consciousness or God is the ground of all phenomena.
5. Our ultimate goal is mystical enlightenment, or union with God, and its pursuit is the greatest good of human existence.[3]

The Perennial Philosophy
of the *Bhagavad-Gita*

The *Bhagavad-Gita*, which means "God's song," is a road-map for living life in peace, harmony, and love. This epic poem of ancient India, whose author is generally considered to be Vyasa, who lived between the fifth and third centuries BCE, is a conversation between Arjuna, a great and noble man, and Krishna (God). Arjuna's and Krishna's discussion takes place in the immediacy and the highly charged setting of a monumental battle that is about to begin. Krishna is depicted as the charioteer for the great warrior Arjuna, who struggles over fighting the enemy because they are his own relatives. His father, grandfather, uncles, cousins, nephews, and other relatives have amassed a huge army of thousands to fight against Arjuna and his brothers. The odds for victory are heavily against Arjuna, but he has something his enemies lack: the council of Krishna. Though the battle about to begin is depicted as real in the *Gita*, it is a metaphor for the *internal* battle of human experience.

The reader quickly identifies with Arjuna's plight. The enemy's immense army of thousands of horses, chariots, bows and arrows, and other instruments of war represent many of life's overwhelming challenges. The chaos and dangers of war are metaphorical representations of our mental, emotional, physical, and spiritual needs, often fraught with uncertainties, limitations, confusion, and moral and ethical dilemmas.

But before we address Arjuna's despondency and Krishna's counsel, let's ask this question: Why should we pay attention to anything in the *Gita* at all? Because scholars, philosophers, and sages around the world have considered the *Gita* one of the greatest sources of wisdom, inspiration, and truth. Countless people through the ages have found meaning in this long poem.

Ralph Waldo Emerson mentions the *Gita* often in his *Journals* with reverence: "It was the first of books; it was as if an empire spake to us, nothing small or unworthy but large, serene, consistent, the voice of an old intelligence which in another age & climate had pondered & just disposed of the same questions which exercise us."[4] Henry David Thoreau considered the *Gita* a masterpiece, and spoke of it in glowing terms: "In the morning I bathe my intellect in the stupendous and cosmological philosophy of the *Bhagavad Gita,* since whose composition years of the gods have elapsed and in comparison with which our modern world and its literature seems puny and trivial."[5]

The relevance, beauty, and wisdom of the *Gita* noted by two American sages gives it unique credence and relevance for the contemporary reader. But the relevance and significance of any work of art or literature shouldn't be based solely on the opinions of well-known people or authority figures. The value and importance of the *Gita* is best determined in the heart of one who finds it.

The Gita and I

The recipient of my adoration seemed to have it all. Beauty, charm, strength, determination, intelligence, humility, and a desire to learn, grow, and live well.

What first caught my attention was her ability to hold the most difficult yoga postures—she was better at it than the instructor leading our yoga class. When we first spoke, I noticed the ease and softness of her voice, as well as her curiosity about me. Later she told me, "I was impressed by your self-confidence." I hadn't considered this trait to be particularly strong in me, but her remark made me feel good. I wondered if she would like me and felt it would be easy for me to fall in love with her.

Soon we became lovers. It was incredibly exciting for me to be with her, but it wasn't long before I realized that my passion, desire, and attraction were misguided. I knew that our relationship was not sustainable, but emotionally it was extremely difficult for me to end it. Each time I attempted to move on, I failed.

One day during a discussion of the *Bhagavad-Gita* at a meditation center, I finally found a sense of peace and the conviction to end the relationship, a feeling that emerged unexpectedly. *How did it happen?* I wondered. Was I simply ready to hear the message of the *Gita* in that moment? Did the peaceful energy I always felt at the meditation center have something to do with my sudden realization? Or was it Krishna's counsel to Arjuna that made me feel this way? Perhaps it was a mixture of everything at once. But what stood out for me the most was the immediate kinship I felt with Arjuna. I resonated with his despondency, as well as Krishna's counsel, and it seemed as if Krishna was speaking directly to me.

Prior to reading the *Gita* I had spoken to my sisters and a couple of friends about my difficulty in ending the relationship. Everyone was encouraging and supportive, but I couldn't act on their advice, nor did I want to. It wasn't until I realized that my despondency and my inability to act were similar to Arjuna's that I understood I was not alone in feeling this way. Krishna's advice lifted my burden of feeling weak and flawed, and gave me the courage to make appropriate choices.

It wasn't easy or immediate, but it set me on the path to recovery. The sense of confidence and peace I gained permitted me to reflect, contemplate, meditate, surround myself with sympathetic friends, learn about energetic healing and chakra alignment, spend time in nature, and pray ceaselessly. Eventually, a combination of all of these practices helped me recover.

Still, if I had not had a vivid dream one night, I would have struggled much longer. We will get to that dream in a bit.

The sudden and unexpected awareness that triggered the beginning of my healing was a shift in my consciousness. It didn't seem like something I had done on my own. Rather, it felt like something inside of me had been awakened. William James, considered the founder of American psychology, described in his classic work *The Varieties of Religious Experience* (1902) how human consciousness transforms. He explained two forms of change: The first is gradual and continuous, like the opening of a flower. The second is sudden or abrupt. In the latter case, change is often associated with what James calls "mystical states of consciousness."

A scientific explanation for my sudden and unexpected healing is best explained in Deepak Chopra's book *Quantum Healing: Exploring the Frontiers of Mind/Body Medicine*. Chopra suggests that what we think and believe has a profound influence on all aspects of our health and well-being. Though Chopra focuses on spontaneous or quantum healing from the perspective of psychoendoneuroimmunology, his implicit explanations convey the role of consciousness, which he suggests is behind all types of healing.

My recovery was both sudden and continuous. While discussing the *Gita*, I suddenly realized that I could handle my pain, yet the healing process and the actions I took came gradually. Part of my healing was realizing that everything I experienced was meant to be—not as in destiny, but rather as in the framework of the Great Mystery. I couldn't understand what happened to me through binary logic, though I gained a "sense" of it in the framework of tetralemma logic.

Some Key Points on the *Gita*

As I indicated earlier, I am not advocating the wisdom of the *Gita* from a religious perspective. Nor are the insights I present here solely mine. I have drawn from interpretations and translations of the great Mahatma Gandhi; Sanskrit scholars Eknath Easwaran, Barbara Stoler Miller, and Swami Rama; and noted author Stephen Mitchell. The philosophical, nonlinear, paradoxical, and intentionally multifaceted teachings of this epic poem allow all of us to find our own meanings and understandings.

However, it is difficult to miss the key insights of the *Gita*. The importance of meditation, yoga, reflection, contemplation, and knowing the nature of reality and who we are is repeated throughout the poem. Krishna states that not only do our senses create a false understanding of reality, but they are also extremely compelling. To get past these false understandings requires us to practice techniques to steady the mind. Today, with increasing awareness of meditation and yoga, it may be easier for us to pay heed to this ancient wisdom than it once was.

Let's return to Arjuna and the battleground. The sounds of conches, the neighing of horses, and the din of the enemy warriors are ominous. The chaos and violence of impending war is moments away, yet Arjuna cannot bring himself to fight his relatives who have raised him, taught him about life, and been his dear companions. Dejected and filled with pity he says,

"Krishna I see my kinsman
gathered here, wanting war.

My limbs sink,
my mouth is parched,
my body trembles,
the hair bristles on my flesh.

The magic bow slips
from my hand, my skin burns,
I cannot stand still,
my mind reels.

I see omens of chaos,
Krishna; I see no good
in killing my kinsman
in battle."6

Arjuna emphatically says to Krishna, "I will not fight," and falls silent. Then, to alleviate Arjuna's despondency, Krishna responds by stating that life is eternal, with no beginnings and endings, and therefore Arjuna must engage in battle and not be disheartened by thoughts of killing and death. Because life is eternal, death is an illusion. Krishna informs Arjuna that upon physical death, everyone and everything reunites with him, bringing everlasting peace, love, and infinite knowing that are not possible in physical life. The metaphors of war and death are not meant to suggest violence, murder, or vanquishing the enemy; instead, they convey deeper truths about life and the nature of reality. As Arjuna sits feeling downcast, dejected, and helpless, Krishna smiles at him and speaks these words.

"Although you mean well, Arjuna
Your sorrow is sheer delusion.
Wise men do not grieve
*for the dead or for the living."*7

"Never have I not existed,
nor you, nor these kings;
and never in the future
shall we cease to exist.

Contacts with matter make us feel
heat and cold, pleasure and pain.
Arjuna, you must learn to endure
fleeting things—they come and go!

When these cannot torment a man,
when suffering and joy are equal
for him who has courage,
he is fit for immortality."[8]

Let's review the practices and understandings that lifted Arjuna from his despondency. His pain, confusion, and inability to act are similar to the sorrows, addictions, and problems of life. And it is possible to not only recover from them, but also to experience equanimity when these are acted upon in accordance with Krishna's counsel. Though similar messages exist in the religions of the world—as well as in the philosophies, poetry, and literature of modern and ancient times—I was reading the *Bhagavad-Gita* when I needed its counsel the most. My insights could have come in a church, mosque, or synagogue. They could have come to me while I was attending a class, speaking to a friend, or listening to an audiobook on how to get over a breakup. The point is that God speaks to us, telling us just what we need to hear when we need it most, in myriad ways.

The *Gita* is a map to help us negotiate life's complexities. Its wisdom for resolving the conflicts of our minds and hearts is penetrating, reaching our sensibilities and speaking to our soul. Krishna counsels Arjuna on the issues of wisdom and ignorance; right and wrong; courage and cowardice; mortality and immortality; self-knowledge, sacrifice, and renunciation. As we read the verses of the *Gita*, Krishna reveals himself in stunning and vivid detail by giving Arjuna a divine eye. Upon seeing the true and infinite nature of Krishna, Arjuna says,

> *"I see you blazing*
> *through the fiery rays*
> *of your crown, mace, and discus,*
> *hard to behold*
> *in the burning light*
> *of fire and sun*
> *that surrounds*
> *your measureless presence.*
>
> *You are to be known*
> *as supreme eternity the deepest treasure*
> *of all that is,*
> *the immutable guardian*
> *of enduring sacred duty;*
> *I think you are*
> *man's timeless spirit.*
>
> *I see no beginning*
> *or middle or end to you;*
> *only boundless strength*
> *in your endless arms,*
> *the moon and sun in your eyes,*

your mouths of consuming flames,
your own brilliance
scorching this universe."[9]

In seeing Krishna in his infinite forms, Arjuna realizes that he, too, is an intricate part of the universe, and that in seeing himself in the image of Krishna, he will triumph: triumph over his self-delusions and the mind-created limitations he has internalized over his lifetime. It will take courage and faith for him to gain knowledge and to act in accordance with his magnificence and true nature.

The significance of the *Gita* for me was in realizing that at a deeper level, everything, every action, and all that we experience or know are direct representations of God, and that through *meditation* it is possible to experience God's peace and presence. The importance of meditation in achieving peace or enlightenment is delineated in the *Yoga Sutras of Patanjali*, which we will discuss next. But before we do, it is important to briefly touch on the meaning and spirit I associate with God.

Belief in the existence of something infinitely greater than ourselves has been around for as long as humans have walked on this earth. It has been referred to as God, Allah, Jehovah, Brahman, Buddha, the Lord, Universe, Great Mystery, Source, Divine Intelligence, and Self, among other names. The terms used are culturally bound and represent philosophical or religious beliefs, but these references to something much greater are also akin to what science has discovered.

In a larger sense the need to believe in something greater is a human tendency, and therefore all faiths, rituals, and ceremonies have meaning and purpose for their respective practitioners. As I reflected upon the perennial philosophy and the scientific understandings of the nature of reality, I allowed the concept of an infinitely great power beyond understanding and rational thought to enter my heart, and I understood it to be God.

The God I speak of is not the God of any religion. God is beyond description or conceptualization. Beginnings and endings are not in the realm of God. Creation and destruction are part of God, but God is even beyond these phenomena. God cannot be named, for a name signifies a conceptual reality. Not knowable, yet knowable; not real and yet real; beyond identity and yet a presence that can be felt and experienced. This is the God I pray to.

In recognizing this God within me, I recognize myself. In loving God, I love myself. And in knowing God, I know myself. My conscience is a representation of the God in me, as it is in everyone. When I let my conscience guide my thoughts, words, and deeds, I am peaceful. When I let my ego run my life I suffer.

My devotional prayers to God, my praise, appreciation, and gratitude, are not for God's sake, but for mine. The infinite God does not require praise, worship, or devotion. Praise, worship, and devotion to God are for my benefit. And in recognizing God this way, I gain a sense of peace, strength, and harmony that is not accessible elsewhere.

The Perennial Philosophy of the Yoga Sutras

The Yoga Sutras of Patanjali is a work that has been studied and commented upon for hundreds of years—a testament to Patanjali's skill and to the wisdom of yoga's oral tradition. It consists of four short books called *padas* or chapters, which first appeared in Sanskrit. Translations into English have been made by dozens of scholars, pundits, and yogis, with varying opinions on how to interpret the original text.

The Sanskrit word *sutra* is akin to *suture* or thread. This term accurately captures Patanjali's intent and effort to thread the oral history of yoga into a coherent tapestry.

The four chapters of the *Yoga Sutras* are:

1. *Samadhi Pada*, a list of fifty-one aphorisms about the nature of meditative absorption
2. *Sadhana Pada*, fifty-five aphorisms on the practice of yoga
3. *Vibhuti Pada*, fifty-six aphorisms on the *siddhis*—the extraordinary capacities (ESP and other supernormal abilities)—that one learns on the yogic path
4. *Kaivalya Pada*, thirty-four aphorisms about the goal of yoga: liberation or enlightenment.

In essence, the *Yoga Sutras* teach that to achieve enlightenment one must master two principle skills: the first is dispassion (somewhat akin to "attached-detachment" in

Buddhism), and the second is absorption through a deep meditative state called *samadhi*.[10]

The *Yoga Sutras*

It is true we are born without an instruction manual for life, but instructions for living a successful and happy life have been around for a long time. Patanjali's *Yoga Sutras*, written more than two thousand years ago and an integral part of the perineal philosophy, is a comprehensive delineation of who we are and our ultimate purpose. In studying the 196 sutras or aphorisms of this text, we find meaning, purpose, and happiness, regardless of our circumstances in life. Patanjali's central message is that we are surrounded in every respect by the intrinsic power and creativity of the Divine. At our core we are infinitely pure, luminous, loving, and wise. "Discovering the intrinsic divinity at our core is the highest achievement; dying without experiencing it is the greatest loss."[11]

The first lesson of the *Yoga Sutras* is to practice meditation, which leads to the direct experience of our inherent grandeur. The clarity, confidence, and focus we experience in the regular practice of meditation moves us toward fulfilling our life's purpose. Without meditation we lack peace of mind, which blocks our ultimate purpose. This occurs because "the mind has two attributes: one innate and the other acquired."[12] In chapter 5 we identified these two attributes as Self-1 and Self-2, which we can now discuss in the context of the *Yoga Sutras*.

Attributes of the Mind

Self-2 or the infinite, luminous mind is an innate attribute. This self-contained and self-guided inner radiance manifests as genius and empowers us to excel in life. It enables us to unravel the mysteries of the forces of nature, and it is also the ground for spiritual revelations. This Higher Mind is limitless. "When directed outward it unveils the mysteries pertaining to the external world: we become scientists. When directed inward, it unveils its own mysteries and sees the Seer: we become sages."[13]

Self-1 coexists with Self-2 but is unaware of such a relationship. This lack of awareness causes Self-1 to create a convoluted version of reality, and then Self-1 works hard to protect it. Pandit Rajmani Tigunait states:

It gives rise to distorted self-identity, attachment, aversion, and fear of loss. These acquired attributes veil our inner luminosity, forcing us to grope our way in the darkness of ignorance. We all are trying to discover the secret of life. We want to know who we are, what our relationship is with the world, what is the true source of happiness, and most importantly, how to live with purpose and without fear. We know the mind has enabled us to master the external world, but we rarely stop to consider that if we master the mind itself, and thereby gain access to its vast pool of power and intelligence, unending joy and eternal freedom will be ours. Instead, we let the mind roam aimlessly. Protecting it from doubt, fear, anger, confusion, and self-incrimination is not among our priorities. As a result, we strive for peace with an agitated mind. We look for clarity with a stupefied mind. We search for our inner self with a mind that knows only how to operate in the external world. We attempt to achieve lasting happiness with a mind accustomed to chasing short-lived pleasure. We yearn for ultimate freedom with a mind enslaved by its own dysfunctional habits. In short, we have set out on the path of conquest with a self-defeating mind. Doubt, fear, uncertainty, sorrow, and grief thus become an integral part of our destiny.[14]

Tigunait adds, "The Yoga Sutra tells us step-by-step how to eliminate these acquired and self-defeating properties and reclaim our innate luminosity."[15]

Karma Chakra

According to Patanjali, the unlocking of the power of our mind requires that we understand the law of karma, which is similar though not identical to "As you sow, so shall you reap." Pleasant and unpleasant experiences enter our lives unannounced, but we don't normally understand what propels these events. The law of karma requires that we fully understand the consequences and relationships of our daily experiences based on our conscience. This realization means understanding that our thoughts, words, and actions leave impressions on our minds. These impressions accumulate and become the drivers of our thoughts and actions, known as *samskaras*.

Furthermore, Patanjali asserts that past karmic impressions also influence our present thoughts and actions. This assertion takes for granted the reality of past lives and therefore differs from the adage "As you sow, so shall you reap," in that the latter is a statement made in the belief that there is only one life. Patanjali explains that if karmic impressions of our past lives are not checked in time, they will lead to deeper karmic impressions (*samskaras*), which then compel us to perform similar actions, creating the cyclic karma chakra.

Over time *samskaras* become so strong that they influence our power of discernment from deep within. Judgment of right and wrong, our perception of ourselves and others, and how and what we think are no longer our own. The mind becomes a slave to its own creation and we are unable to see our core being, which is pure consciousness. Restoring the pristine nature of our mind, in essence, requires the daily practice of meditation or deep absorption, which is the subject of Sadhana Pada the second chapter of the *Yoga Sutras*.[16]

Desires in Yoga

In Sadhana Pada, Patanjali discusses the ways in which human desires, which are as many as the stars in the sky, affect our lives. He divides the entire range of desires into three categories: *sattvic*, *rajasic*, and *tamasic*. He explains that *sattvic* desires are illuminating. In these desires we experience enthusiasm, inspiration, and the will to discover and embrace our purest and highest self. *Sattvic* desires help us change and grow. They help us find the courage to pursue our deepest and most noble convictions.

Rajasic desires leave us perpetually dissatisfied. They make us unsure of ourselves, and we remain confused. *Rajasic* desires make the demands of life seem burdensome, and they sap our enthusiasm, courage, and freedom. In their presence we confuse pleasures with joy. We are unable to appreciate what we have, and we chase what we don't have. Dissatisfied and unhappy, we seek refuge in being busy—but we don't really know why.

Tamasic desires make us lethargic and disinterested. We don't strive to learn, improve, or contribute to the greater good. We lack enthusiasm, and life seems to be a series of listless "same old, same old."

According to Patanjali, the minds of ordinary people are dominated by the interplay of these three categories of desires. When we learn to free our minds from this interplay and learn to embrace only illuminating ones, we become exceptional. Knowing that human beings are prone to having all three types of desires, albeit to varying degrees, is helpful. It tells us we are not alone in having contradictory desires, and knowing this may prompt us to exclusively make *sattvic* choices. When we make only *sattvic* choices, our path to enlightenment becomes clear and attainable.[17]

Suffering

Patanjali emphasizes the need for purifying the mind by embracing *sattvic* desires exclusively, which is accomplished by turning the mind inward where it has the opportunity to bathe in the luminosity of the Divine. However, he clearly sees that most human beings are caught in the day-to-day struggles of life, and that sorrow, pain, fear, and grief prevent us from the pursuit of enlightenment. He reminds us of our entrenched habits of denying that we are engulfed in sorrow, known as *duhkha* in Sanskrit. This broad term includes the familiar forms of physical pain as well as fear, grief, depression, dementia, and every other condition that causes us to experience life as burdensome. We shy away from exploring the deeper dimension of pain and its crippling effect on life—yet this exploration is necessary. In essence, Patanjali's succinct plan to transcend suffering and embrace lasting peace happens when we make meditation and yoga central aspects of life. He reminds us that peace and happiness are intrinsic qualities of these practices. No particular faith or belief is necessary for their achievement.

One of the most frequent causes of suffering is our incessant internal chatter. We review the various ways in which our feelings were hurt. We engage in mental dialogues with people we perceive have spoken harshly to us or insulted us. We try to set them straight by silently telling them how unfair and hurtful they were. These imagined dialogues plague our minds, creating stress hormones in our bodies. Constant worry, anger, fear, and other negative emotions damage brain cells and often result in serious illnesses.

One way to alleviate suffering is to be mindful of the internal chatter and to stop it. Letting go of the agitated "monkey mind" that jumps from one thought to another may not be easy, but it is possible when we are grateful for even our most ordinary capacities: gratitude is a powerful antidote to suffering. For example, we normally take our abilities to see, taste, smell, touch, and hear for granted, but expressing gratitude for them helps shift our attention toward positive thoughts and emotions.

Whenever I catch myself in meaningless internal chatter, I say to myself, *There you go again! Do you realize how stupid you would sound if you spoke your thoughts out loud?* Imagining my thoughts being said aloud makes me laugh, and I let go.

We can also shift to positive thoughts and feelings by focusing on a poem or a passage of prose that expresses hope, love, generosity, and compassion. Affirmations, chanting, silence, and triggering the relaxation response are some other ways to alleviate pain and sorrow. Quiescence and bliss are our fundamental nature, as is reflected in oft expressed themes from Hindu and Christian scripture: *Atman and Brahman are one* and *The Kingdom of Heaven resides within us.* To overcome suffering and achieve enlightenment, Patanjali describes a method known as *ashtanga,* or the eightfold path.

The Eightfold Path

Path 1 is *Yama*, restraining and avoiding harmful behavior toward self and others. This includes violence, injury, lying, stealing, sexual exploitation, greed, and coveting the possession of others.

Path 2 is *Niyama*, to deal more internally with one's own practices, but practices still related to external elements—hygiene, austerity, contentment, disciplined practice, devotion, and self-study.

Path 3 is *Asana*, the development of physical postures to assist the mind and body in relaxing, through development of strength, steadiness, and flexibility. The purpose of the asanas is to prepare the body to comfortably withstand the rigors of long-term meditation.

Path 4 *is Pranayama*, conscious breathing techniques, which further the mind's ability to focus and energize the body.

Path 5, *Pratyahara* continues the progression of internalization by withdrawing consciousness itself from the senses, fostering even greater tranquility of mind.

Path 6 is *Dharana*, developing a steady, sustained concentration by focusing attention on any external object as a place upon which the mind can be fixed. This type of concentration is similar to that experienced during highly focused intellectual work.

Path 7 is *Dhyana*: developing prolonged levels of concentration on an object, with deeper absorption and greater sustained alertness, referred to as meditation.

Path 8 is *Samadhi*, unity or mystical absorption with

an object of attention. In this state, distinctions between subject and object dissolve and one "becomes" the object of meditation. This awareness is often described as ecstatic.[18]

When one follows the eightfold path, not only is suffering alleviated, but one can gain enlightenment, as well as "superhuman" abilities known as *siddhis*. In the third chapter of the *Yoga Sutras*, Vibhuti Pada, Patanjali describes these *siddhis*, which we will discuss later. But because most of us are caught in the day-to-day struggles of life, enlightenment remains a far-fetched dream. Therefore, first we will review some "down to earth" reasons for practicing meditation, as well as some other ways to remain grounded and at peace.

Meditation

Why meditate? Life is stressful, and in the day-to-day hustle and bustle it's easy to get wrapped up in negative thoughts and feelings of anxiety, anger, and frustration. We may want to achieve peace of mind, but we don't know how. In not knowing that it's possible to remain at ease even under stressful circumstances, we tend to give up and become apathetic or even depressed.

Most of us, from time to time, experience moments of inner peace, altruistic love, and compassion for others, but for the most part these are fleeting. What if it were possible for us to cultivate these feelings and make them permanent? No doubt it would be preferable to experiencing the pull and tug of negative emotions and painful memories. Wouldn't it be much better to have inner fulfillment, and contribute to the well-being of others? Remarkable as it may seem, the practice of meditation makes all of this possible.

A significant reason for meditating is to become a better human being. Most of us believe that we are who we are. We are born with certain characteristics, and essentially it isn't possible for us to change them. For example, loving-kindness, compassion, generosity, and our ability to forgive others may be thought of as fixed qualities of our nature and temperament. Though we may feel that we could be better at any and all of these qualities, we normally don't take the time to learn or develop greater compassion or loving-kindness. We may not know that meditating transforms our brains for the better. We may not know that meditation makes us more compassionate, loving, appreciative, and forgiving. Our sense of well-being and our capacity to navigate life's ups and downs become easier. We become more tolerant, less reactive, and better able to live with greater ease.

Because many of the general health benefits of meditation, such as stress reduction, have been covered in books, magazines, and newspapers, most of us are aware of them. However, most people may not be aware of the scientific studies that substantiate the extraordinary capacities of meditation to fundamentally alter the brain for the better.

Sara Lazar, a neuroscientist at Massachusetts General Hospital and Harvard Medical School, found that in comparison to a control group, seasoned meditators had increased amounts of gray matter in the insula and sensory regions of the brain, the auditory and sensory cortex. Her research also documented more gray matter in the frontal cortex, which is associated with working memory and executive decision making. Since the cortex shrinks as we get older, an interesting finding was that the prefrontal cortex of fifty-year-old meditators had the same gray matter as twenty-five-year-olds.[19]

Lazar also found differences in brain volumes in five different regions of the brain after an eight-week period of meditation. In the group that learned meditation, she found thickening in four regions:

The posterior cingulate, which is involved in mind wandering and self-relevance.

The left hippocampus, which assists in learning, cognition, memory, and emotional regulation.

The temporoparietal junction, which is associated with gaining perspective, empathy, and compassion.

An area of the brain stem called the pons, where many regulatory neurotransmitters are produced.

Meanwhile, the amygdala—the fight-or-flight part of the brain where anxiety, fear, and stress generally register—got smaller.[20]

In a paper published in the *Indian Journal of Psychiatry* (2009), Chittaranjan and Radakrishnan document several studies conducted all over the world, writing,

Meditation has been found to produce a clinically significant reduction in resting as well as ambulatory blood pressure, to reduce heart rate, to result in cardiorespiratory synchronization, to alter levels of melatonin and serotonin, to suppress corticostriatal glutamatergic neurotransmission, to boost the immune response, to decrease the levels of reactive oxygen species as measured by ultraweak photon emission, to reduce stress and promote positive mood states, to reduce anxiety and pain and enhance self-esteem, and to have a favorable influence on overall and spiritual quality of life in the late-stage disease. Interestingly, spiritual meditation has been found to be superior to secular meditation and relaxation in terms of decrease in anxiety and improvement in positive mood, spiritual health, spiritual experiences, and tolerances to pain.[21]

How does meditation help us learn, think, and feel better and improve our health? How and why do we experience greater peace of mind, compassion, empathy, and well-being simply by meditating? Science provides measurable and observational answers to these questions, but deeper answers are found in spirituality. From a spiritual perspective, our bodies, brains, hearts, and souls are connected to the Source through which we receive love, wisdom, and transformation. Thus, meditation is the process of creating a deeper connection with God, in whose presence we are transformed for the better. Another significant aspect of meditation is that irrespective of one's faith, belief, or nonbelief (atheism), anyone who meditates regularly experiences peace and transformation. With so many benefits to meditation, it makes sense to learn how it's done.

The Practice of Meditation

The benefits of meditation are not dependent on any faith, religion, or outlook. Simply being aware that something much greater than your own thoughts and feelings is at work can lead to contentment and peace. There are many forms of meditation, and details about them are readily available in bookstores and online. In this section we will look at three kinds of meditation common in Buddhism, which today are practiced in secular settings like hospitals, schools, and even in the corporate world.

1. *Focused-attention* meditation, aiming to center the mind and develop the capacity to remain vigilant to distractions.
2. *Mindful* meditation, meant to cultivate awareness of stressful or difficult emotions, thoughts, and sensations of the present moment.
3. Meditation to develop *compassion* and *loving-kindness* for one's self and others.

There are some common practices for all three forms of meditation. In his book *Why Meditate? Working with Thoughts and Emotions,* Buddhist monk Matthieu Ricard, a spokesperson for the Dalai Lama, gives a good description of these three types of meditation. In the following, I have summarized a few details from Ricard's book and integrated them with my own meditation practice.

How to Meditate?

First, it is best to find a competent teacher. A knowledge-able teacher can be a source of guidance and inspiration, but in many areas such a person may be difficult to find. However, almost all major cities in the US have spiritual or meditation centers where competent teachers are available to help. And there are hundreds of quality videos and guided meditation exercises on the Internet. It's a matter of finding one that suits you.

Next, it is important to have a suitable place and setting. A quiet place to meditate while wearing loose clothing is appropriate. It's best to set aside ten or twenty minutes without interruption in the beginning. You may want to increase the time to forty minutes as you become comfort-able with your practice. But the exact amount of time is not as important as your commitment to meditating. In fact, two-minute meditations in the middle of the day are quite helpful. Feel your way regarding the time you spend in meditation. Twenty minutes in the morning and evening is ideal for many people.

An appropriate physical posture is also important. Sitting cross-legged on the floor, a posture often called the lotus position, is recommended. One foot rests on top of the opposite thigh, and the other foot is placed under the other thigh. It helps to sit in this position with a straight back and with your hands resting palms up on your lap in the posture of equanimity, in which the right hand is on top of the left and the tips of the thumbs touch each other. Some practitioners prefer to have their hands rest with palms flat on their knees.

Notice your shoulders. If one is hunched more than the other, relax and bring the shoulders into balance. Keep your chest open to breathe easily with your chin tucked in slightly toward the throat and the tip of your tongue touching the palate, near the front teeth. Some practitioners prefer to keep their eyes open and focused gently on an object. Others prefer to have their eyes half shut. Most of the people I meditate with at my local meditation center keep their eyes closed, and I do too. Because I'm unable to sit in the lotus position, I sit at the edge of my bed or on a chair with my feet flat on the ground. I am not reluctant to modify my posture when it becomes uncomfortable, without being fidgety.

Once you are situated, the practice of meditation is easy and straightforward. I start with the intention to steady my mind. I focus on my breath, in and out, while breathing through my nose, and say a mantra silently (the mantra "So Hum"—which translates "I am That"—is one that anyone can use). I have also used the words "I am, at peace" instead of the mantra. These words or variations of them work just as well. On the in breath I say, "I am" and on the out breath I say, "at peace." The rhythm of the words and the intention behind them are what's important.

During meditation various thoughts enter my head, and they can be disruptive. When this happens I simply observe them without attachment or judgement, and I allow them to float by in my mind. I have found that after a while the incessant thoughts seem to diminish. I also make sure that I don't grasp for ease or demand to be at ease. I simply let it happen. In this approach I experience a pleasant and nurturing sense of quiescence.

That's it. That's all there is to it. However, the habit of practicing meditation twice a day requires commitment and discipline. It is normal for most people to waver from their meditation practice from time to time. If this happens to you, don't fret, but gently bring yourself back to your practice. Nothing should feel forced or compulsory. Over the last forty years I have started and stopped meditation many times.

Mindfulness Meditation

We are often distracted, scattered, and confused as our minds remember the past and worry about the future. We miss what is taking place in the moment. We lose sight of the basic awareness that always lies behind these thought processes. The antidote for a stressed and anxious mind is the art of mindfulness. Jon Kabat-Zinn, professor emeritus of medicine and the creator of the Mindfulness-Based Stress Reduction program at the University of Massachusetts Medical School, is a prolific writer on and teacher of mindfulness. In his book *Full Catastrophe Living: Using the Wisdom of Your Body and Mind to Face Stress, Pain, and Illness*, he provides an excellent guide for living mindfully. He explains that mindfulness consists of being fully aware of everything that arises within and around us from moment to moment. Paying attention to and being aware of what we see, hear, feel, and think, along with gaining understanding of the nature of our perceptions, helps us in living well. Zinn's guidelines are useful. However, in essence, practicing mindfulness meditation is straightforward and simple. In fact, mindfulness is our natural state of being. We can experience it just by being present.

A Mindful Moment

To bask in the morning sunshine, I opened the glass door that leads to my deck with its expansive view of the back-yard and surroundings. As I stepped out onto the deck, I felt a cool breeze caress my face. Almost instantly I felt at ease. The sunshine and the freshness of the beautiful morning bathed my senses. The vastness of the blue sky pulled me in. I belonged to this expanse. I heard birds singing in the pines and maples. The lush greenery in contrast to the blue sky had a vividness and brilliance I hadn't noticed before. Red, purple, pink, and white petunias in flower pots glistened in the sunshine, and I couldn't resist touching them. Looking at the sky, touching the petals, listening to the birds, and feeling the morning breeze, I sank into this infinite moment. My body, mind, heart, and soul were one with everything. I couldn't help but lift my arms toward the sky to express gratitude for the awe, wonder, and beauty. It reminded me of what William Blake wrote:

To see a World in a Grain of Sand,
and a Heaven in a Wild Flower,

Hold Infinity in the palm of your hand,
and Eternity in an hour.

It was not easy for me to pull away from this peaceful moment, but my desire to share this moment with you made it easier to get back to work. I also felt that it would be easy to remember the moment for a long time. We have all had such peak experiences, but most of the time we struggle. Another way to remain centered lies in meditating with loving-kindness.

Loving-Kindness Meditation

Almost everyone at some time or another has felt benevolence and altruistic love for those who suffer. However, most of us, most of the time, are focused on our own well-being, and we vary in our compassion toward others. Compassion in Western thought is understood as sympathetic pity and concern for the suffering of others, but from the Buddhist point of view compassion toward ourselves is as important as being compassionate toward others. Furthermore, Buddhism considers cultivating compassion essential for living well, and loving-kindness meditation to be a way to develop it. To this end, Buddhists advise us to cultivate four particular attitudes: altruistic love, compassion, joy in the happiness of others, and equanimity. And these should first be directed toward ourselves.

It helps to begin the loving-kindness meditation with this prayer: "May I be strong, may I be happy, and may I live with ease." Repeat this or a similar prayer for a while. Next, think of someone you love. It could be your son, daughter, sibling, parent, friend, or lover. Imagine benevolence, unconditional love, and perfect well-being for this person. After that, extend loving-kindness toward others like your neighbors, coworkers, or acquaintances. Finally, include this wish for your enemies or the enemies of humanity. This doesn't mean you should condone their actions or their hateful plans. Instead, formulate thoughts and feelings in which you wish for them to give up their negative thoughts and behaviors and instead develop benevolence and care for the happiness of others. Do this knowing that the greater the sickness of another person, the more they need attention and goodwill. For all of the people toward whom you extend thoughts of loving-kindness, you may use this simple Buddhist prayer: "May you be strong, may you be happy, and may you live with ease." I have found that if I name a person in this loving-kindness statement instead of just saying, "you," I can remain engaged with the prayer and meditation with greater ease.

The Power of Prayers

Prayer has been a part of religious practices all over the world for thousands of years, and many people have found comfort, peace, and healing by praying. Yet the efficacy of prayer has not been accepted by mainstream science, even though scientific evidence shows that prayers work. Many double- and triple-blind studies have shown the positive, observable effects of prayers.

The current gold standard for the investigation of the

efficacy of medical interventions is the double-blind, randomized controlled trial. Most recent studies on prayer and healing have adopted this design. Commonly in such studies, intercessory prayers are offered for the health of patients selected at random. These patients do not know they are being prayed for, and the persons who are praying do not come into contact with the patients for whom they pray. Medical outcomes in these patients are compared with outcomes of patients in a control group who are not prayed for. Finally—and importantly—the medical treatment team is also blind to the prayer group status of individual patients. Thus, these studies are triple-blind.

Following is a summary of one such study:

[Kwang Y] Cha *et al* studied 219 consecutive infertile women, aged 26–46 years, who were treated with in vitro fertilization embryo transfer in Seoul, South Korea. These women were randomized into distant prayer and control groups. Prayer was conducted by prayer groups in the USA, Canada and Australia. The patients and their providers were not informed about the intervention. The investigators, and even the statisticians, did not know the group allocations until all the data had been collected. Thus, the study was randomized, triple blind, controlled and prospective in design.

Cha *et al* found that the women who had been prayed for had nearly twice as high a pregnancy rate as those who had not been prayed for. Furthermore, the women who had been prayed for showed a higher implantation rate than those who had not been prayed for. Finally, the benefits of prayer were independent of clinical laboratory providers and clinical variables. Thus, the study showed that distant prayer facilitates implantation and pregnancy.[22]

In another study, K. T. Lesniak described the effects of intercessory prayer on wound healing in a nonhuman primate species. The sample comprised twenty-two bush babies (*Otolemur garnettii*) with wounds resulting from chronic self-injurious behavior. The animals were randomized into prayer and control groups of similar sizes. Prayer was conducted for four weeks. Both groups of bush babies additionally received L-tryptophan. Lesniak found that the prayer group animals had a greater reduction in wound size and a greater improvement in hematological parameters then the control animals. This study is important because it was conducted in a nonhuman species; therefore, the likelihood of a placebo effect was removed.[23]

Many other similar studies have shown that intercessory prayers work. In *God, Faith and Health: Exploring the Spirituality–Healing Connection,* Jeff Levin, PhD, documented the power of prayer and showed that prayers are not dependent on any specific religion. The intention and heartfelt concern for others are all that is needed.

Though there is ample scientific evidence of the efficacy of prayer, skeptics still don't believe in it. Their responses range from outright rejection of the power of prayer to questioning the methodologies and protocol of the studies. But the rejection of skeptics does not take away from the positive values of prayer. Larry Dossey, MD, in his books *Healing Words* and *Prayer Is Good Medicine*, has documented the validity of numerous scientific studies on prayer and suggests using prayer along with the standard practice of medicine involving drugs and surgery. For him it is not an either-or proposition, because there is substantial data showing that prayer helps in the healing process.

I was unaware of the healing power of prayer before I read some of this material, but I found the evidence that Levin, Dossey, and others present so compelling that after encountering it, I concluded anyone with an open mind would believe without hesitation in the efficacy of prayer. As we have discussed, belief is at the heart of the placebo effect, and the extensive research findings of the mind-body connection further signifies the importance of belief in all aspects of health and well-being.

Dean Radin, PhD, of the Institute of Noetic Sciences recorded data collected by computer programs that produce random numbers (called random number generators, or RNGs), which are generally used by actuaries, scientists, and others in their work. The sources of randomness in the RNGs included electronic noise in resistors and quantum tunneling effects in diodes. Radin used 185 RNGs with synchronized clocks, located in different parts of the world (Africa, Asia, Australia, Europe, and North and South America), and their respective data was fed into a central web server in Princeton, New Jersey.

From 1998 to April 2005, Radin and his colleagues documented 185 events when millions of people's attention was focused on the same thing. These included New Year celebrations, natural disasters, massive group meditations, sports events, and outbreaks of war and peace. Examination of the resulting RNG data revealed a change from chance expectation to predictable changes, or *loss of randomness in the data*. It didn't matter what or how people thought or felt about the event, only that their attention was coherently focused on the same one.[24] If human thought can influence machines, it is not too far-fetched to understand that prayers would work as well.

I have found that it's best to pray in a peaceful state of mind reached during meditation. I also believe that prescriptive prayers demanding an outcome don't work too well. Even though I am aware of this, I still catch myself doing just that from time to time. When this happens, I gently observe and shift my prayer to asking what is best for me. I remind myself that I cannot, nor should I, try to control the outcome of my prayers. Taking an approach of "Thy will be done" is best.

Well-Being

We are often unaware of the differences between momentary pleasures and a sense of well-being. We tend to confuse the two, but well-being is a deeper and more lasting state. Momentary pleasures abound. We experience them while listening to music, enjoying a delicious meal, watching an interesting movie, reading a good book, or having sex. Though pleasurable activities are helpful in attaining well-being, they are not necessary, nor do they automatically lead to peace, joy, and happiness. Well-being requires more.

We need to engage in meaningful work and contribute to the greater good. We must feel that our life matters. We need to have good relationships and a sense of belonging to a community. And we must acknowledge and attend to our spiritual selves. But attending to these issues in balance is difficult. Attainment of peace and well-being requires that we develop equanimity through meditation, and the knowledge to alleviate suffering.

Simplicity, as Mahatma Gandhi practiced it, creates peace and well-being, while the desire for more, driven by a sense of lack or insufficiency, creates inner turmoil. To alleviate feelings of disharmony and agitation, we resort to various forms of pleasure, which are momentary and lack equanimity in the long run. Perfectly emulating Mahatma Gandhi's life of simplicity isn't something most of us could do or would even prefer, yet it is entirely possible for us to live a simpler life when we become aware of the turmoil and disharmony associated with seeking peace of mind through pleasures and possessions. Another way to alleviate suffering is the philosophy of *attached-detachment*.

Buddhism's Attached-Detachment

Attachment to any idea, person, or thing makes us feel good, but attachments also have a harmful downside that we don't often realize. We alleviate suffering when we practice attached-detachment. The following story illustrates this principle.

Once upon a time there lived a Buddhist monk a few

miles away from the outskirts of a village. He practiced meditation and lived a simple life, attending to his vegetable garden and tending to a few goats from whose milk he made yogurt and butter for his modest diet.

In the village lived a young man and woman who fell in love. The woman got pregnant out of wedlock. The couple imagined the disapproval and shame the other villagers would heap upon them, and they got scared. So the couple decided to make up a story, blaming the monk for the woman's pregnancy. Upon hearing this, the villagers became angry. They went to the monk and admonished him for his shameful and despicable act. They told him that when the child was born he would have to raise the infant himself. Upon hearing their dismay and anger, the monk gently responded to the angry crowd, "Is that so?"

When the child was born, the monk willingly accepted the responsibility of raising her. But then, after a few years, the now married couple missed their daughter and con-fessed to the villagers that they had lied to them. "Now that we are married we want you to know that we made a mistake, and that we would like our daughter back." The villagers were remorseful. They went to the monk to apologize and request that he return the child to her parents. "We are very sorry for the trouble we created, and the burden of a lie you have had to endure, but we would like for you to return the child."

The monk listened attentively and said, "Is that so?" and returned the little girl to the villagers. The couple was happy and grateful to the monk for having raised their daughter with unconditional love. And the villagers not only learned the folly of being judgmental, but they also gained an abiding respect for the monk.

The moral of the story is evident. We must learn to live with attached-detachment to experience equanimity and peace. We must do our part but not be attached to the fruits of our labor. When we live with attached-detachment, we bring about harmony and change without demanding desired outcomes. This concept was expressed in the *Tao Te Ching*, which translates to "the book of the way," written by the Chinese sage Lao Tzu approximately twenty-five hundred years ago and regarded as one of the greatest books ever written. Lao Tzu's advice of attached-detachment are captured in these words:

> Less and less do you need to force things,
> Until finally you arrive at non-action.
> When nothing is done, nothing is left undone.
> What is a good man but a bad man's teacher?
> What is a bad man but a good man's job?
> If you don't understand this, you will get lost.
> however intelligent you are.
> It is the great secret.[25]

Stephen Mitchell in his translation of the *Tao Te Ching* states:

> Lao Tzu's central figure is a man or a woman whose life is in perfect harmony with the way things are.... The Master has mastered Nature; not in the sense of

conquering it, but of becoming it. In surrendering to the Tao, in giving up all concepts, judgments and desires, her mind has grown naturally compassionate. She finds deep in her own experience the central truths of the art of living, which are paradoxical only on the surface: that the more truly solitary we are, the more compassionate we can be; the more we let go of what we love, the more present our love becomes; the clearer our insight into what is beyond good and evil, the more we can embody the good. Until finally she is able to say, in all humility, "I am the Tao, the Truth, the Life."[26]

In modern life, doing, creating, forcing, expecting, and demanding outcomes is not only considered normal, but it is regarded as reasonable, responsible, and productive way to live and work. Controlling becomes key to progress and the way to a good life. In contrast, the *Tao Te Ching* and the *Bhagavad-Gita* offer a more holistic perspective, where not-knowing, not-doing, not-expecting, and not-demanding are considered equally important. Mitchell's translation of *The Tao Te Ching* states:

Therefore the Master
acts without doing anything
and teaches without saying anything.
Things arise and she lets them come;
Things disappear and she lets them go.
She has but doesn't possess,
acts but doesn't expect.

When her work is done, she forgets it.
That is why it lasts forever.[27]

The Tao is the Absolute, and in harmony with it, all things are possible. In giving up control, we gain. In surrender we are victorious. When we seek nothing, we gain everything. Rudyard Kipling expresses similar ideas in his poem "If":

If you can dream—and not make dreams your master;
If you can think—and not make thoughts your aim,
If you can meet with triumph and disaster
And treat those two impostors just the same;
If you can bear to hear the truth you've spoken
Twisted by knaves to make a trap for fools,
Or watch the things you gave your life to broken,
And stoop and build 'em up with worn-out tools;
Yours is the Earth and everything that's in it ...

Nature has created a state of "not doing" in which all life forms are nourished and nurtured: the magical world of sleep. Let us now examine this wondrous state in which we gain numerous benefits without consciously doing anything.

Sleep

Why do we sleep? The obvious answers are that it restores the energy to function when we are awake, and that it is a natural function of the body. It is akin to hunger, breathing, or digestion. Beyond these commonly understood reasons, researchers over the last twenty years have begun to provide some explanations for why we sleep. But these scientific explanations are in essence limited to the physiological and psychological effects of sleep. They do not really answer *why* we sleep. In the 1990s Jay Allan Hobson, a leading sleep researcher, quipped that the only reason for sleep was to cure sleepiness.

Sleep does not serve just a single purpose. Instead, it is needed for the optimal functioning of a multitude of biological processes. For example, lack of sleep makes the body less sensitive to the hormone insulin, which is produced by the pancreas. And this condition increases the risk of developing type-2 diabetes and obesity. Sleep deprivation impairs learning, memory, emotions, and the immune system. Studies have shown that rats die in just eleven days when deprived of sleep. In fact a study published in 1989 by Carol Everson showed that rats deprived of only rapid eye movement (REM) sleep died in less than two weeks.[28]

This finding is important because we dream during REM sleep, which implies how critically significant dreams are. Why would dreams be so important? Why do we dream? And how should we understand dreams? A discussion about these questions follows, but for now let's continue to focus on the benefits of sleep.

Scientific studies have identified many reasons why sleep is good for us. Sleep apnea, a disorder in which the flow of air into the lungs is interrupted during sleep, causes individuals with this disorder to wake up every few minutes. A 2012 study by the US Centers for Disease Control and

Prevention found that men and women with sleep apnea are, respectively, 2.4 and 5.2 times more likely to have major depression when compared to those without the disorder who were better rested.[29]

Indeed, several studies over the past twenty-five years have concluded that significant lack of sleep can lead to depression and may cause other psychiatric disorders as well. Taken together, the latest research suggests that skimping on sleep affects our hormonal, immunological, and memory functions. If we don't get enough sleep, we can wind up sick, overweight, forgetful, or depressed.[30]

Why we sleep cannot be fully explained in the context of scientific materialism, but from a spiritual perspective we sleep because it transports us into the magical domain of consciousness. In the vastness and serenity of this domain, sleep is freeing, and it resuscitates our spirit by giving us room to roam. Sleep helps us transcend the ordinary physical domain of responsibilities and deadlines, and allows us to experience a larger reality, which holds insights about living well. No one has to instruct us on how to sleep or dream, and yet everyone does these things. The natural healing that takes place during sleep is what sustains us. We come into harmony with what has been referred to as "circadian rhythms," which are associated with the cyclic movements of the Earth and the moon.

Sleeping Patterns and Practices

Sleep is directly influenced by circadian rhythms, so it is therefore prudent to follow their timeline. For example, around 3 p.m., our circadian rhythms demand sleep. If we nap, we feel rejuvenated upon waking. And being in sync with circadian rhythms also requires an early-to-bed early-to-rise practice.

Furthermore, it's best not to watch television, use the computer, or read emails on cell phones prior to falling asleep. The use of these devices interferes with the production of hormones such as melatonin that promote sleep.

A dark room helps us fall asleep. I have two types of blinds on my bedroom windows: decorative blinds that move up and down, and a roll-down flexible blind. When both blinds are pulled down and the doors to my bedroom are closed, the room is completely dark and this helps me sleep well.

Prior to falling asleep, I read for a while, and then review the things I did well that day. I thank God for the good things I did and invite more goodness into my life for the next day. I reflect on things I didn't do as well and pray that I will have greater discernment to do them better the next day. I thank God and ask for dreams that may guide me or inform me of whatever I need to know or do differently in life.

Dreams

One of the most significant reasons for getting a good night's sleep is that it allows us to dream. Dreams are magical indeed! They don't follow the logic of classical physics; they operate in the quantum domain of possibilities. Dreams are not restricted by linear time, nor do they operate in binary logic. In dreams we can fly over canyons and oceans, be in the presence of greats, and even converse with our departed loved ones.

Once during a particularly difficult time of deadlines and responsibilities, I was overwhelmed and unhappy with the monotony and predictability of work and family obligations. Then one night I had a dream. I saw Ammi (my mom) hanging clothes to dry on the line behind our house. She was singing, and she seemed much younger than she was when I had last seen her. In the dream she looked like pictures of her when she was in her late twenties and early thirties, very pretty. The song she was singing was not one I had heard, but its melody was captivating, and I said to myself, "My God, Ammi's voice is beautiful. I didn't know she could sing so well." Even though I didn't know the song, I wanted to sing along. And as I tried to sing, the words came to me. It was an enchanted moment in which I felt serene and joyous.

The message of this dream was gentle, loving, and straightforward. My mother was doing her work of hanging clothes to dry, but while she worked she was peaceful and happy. I felt her love in the words of her song and in her voice. It was obvious to me that she was telling me to live life without worry, anxiety, or fear. I was grateful for the message and found myself working with greater ease in the days that followed.

In dreams we experience a new sense about our thoughts, feelings, and actions. We resolve issues without even trying to. "Sleep on it" is common advice for arriving at difficult decisions, and most of us have at one time or another experienced this. In dreams we resolve conflicts and discharge negativity. In Native American and aboriginal cultures, dreams are considered so important that tribal laws are based on them. Kalahari Bushmen have a saying: "There is a dream, and it is dreaming us."[31] While poetic, this sentiment conveys concepts of quantum theory perhaps better than the descriptions we receive from quantum physicists.

Remembering our dreams and writing them down in as much detail as possible helps capture metaphorically embedded wisdom from the timeless dimension where dreams originate.

Earlier in this chapter I mentioned that a dream I had had helped me end a relationship. In this dream I saw that I was driving my red convertible sports car fast across a long and very high bridge over a river. It was dark, and going fast felt both exhilarating and scary. Unknowingly, I drifted, and the right tire hit the curb, the jolt pumping adrenaline into my system. I felt like I was going to lose control, but I didn't, and I was relieved that the car was running smoothly. Suddenly, while traveling at 75 or 80 miles per hour, I saw that the bridge ended abruptly. I had no time to stop, and I found myself and the car falling into the river below. The height of this incomplete bridge was much greater than the bridge across the Mississippi River that connects Minnesota and Wisconsin, on which I had traveled many times. As I was falling, I thought, *This is it! I'm going to die!* I woke up terrified.

It didn't take but a few moments for me to understand what the dream meant. Hitting the curb and continuing to drive reminded me of having ignored the thoughts and feelings I'd had about ending the relationship. The unexpected ending of the bridge told me that if I didn't end the relationship, it would ruin me. I took heed and soon thereafter took steps to end it.

Some dreams serve to help us cleanse our day-to-day problems and stresses. Such dreams can be scary or pleasant, but they don't require that we do anything about them. We can write them off by saying, "It was just a dream." But remembering our dreams and capturing them in a journal helps us sort through life's predicaments by paying attention to the guidance and predictions embedded in them.

There are many useful books about dreams and how to interpret them. And it helps to read the interpretations and meanings given to dreams by such greats as Carl Jung. But we are the best judges of what our dreams mean. It doesn't matter if we have a nightmare or a dream that makes no sense; we should be grateful for all dreams. If we have this attitude, I believe we can trigger the possibility of more dreams to help us navigate life's challenges and live with greater ease and equanimity.

Siddhis or Superhuman Powers

Let's return now to the discussion of the *siddhis*. The *Yoga Sutras* and Indic tradition do not encourage learning to attain extraordinary psychic powers. But their attainment, though rare, is seen as normal for enlightened beings, who achieve it by gaining proficiency in a refined meditative state known as *samyama*. *Samyama* involves combining expertise in the last three steps of the eightfold path; *dharana*, *dhyana*, and *samadhi*. In the *Yoga Sutras* Patanjali describes numerous superhuman capacities that may be attained while one is on the path to enlightenment. However, he doesn't comment extensively on the latent psychic abilities almost all of us possess, which manifest themselves from time to time without any conscious effort on our part.

These dormant psychic powers sometimes become active in dire circumstances. For example, many people experience knowing of the death of a family member or close friend from afar before being informed about it. It is also not unusual for emotionally connected people to know that a loved one is in danger, regardless of the distance between them or any direct way of knowing about their peril. Furthermore, hundreds of people all over the world have unique and sometimes multiple psychic powers, some of which we discussed in earlier chapters. Though Patanjali doesn't say much about our latent psychic powers, or about the superhuman powers of individuals who are not necessarily on the path of enlightenment, his aphorisms about psychic powers are consistent with present-day observations of psi abilities.

Knowing the Future

When my son Daniel was about a year old, my then wife would pick him up at daycare after work. This facility had a strict policy of closing at 6 p.m., so she was often in a rush to pick him up before it closed. For my part, I worked in Pine City, approximately forty miles away from our house in Forest Lake, Minnesota, and didn't get home from work until 6:30 or 7:00 p.m.

One evening on my way back from work as I neared my house, I had an odd feeling that something wasn't right. As I pulled into my driveway, the feeling got stronger. I was certain something bad had happened. I got out of my car and, as I walked up to the front door, I had a very clear thought that as soon as I walked inside the telephone would ring and a woman would inform me my wife had been in a serious car accident. She would also say that my wife was being transported to a hospital in Saint Paul that was better equipped to care for trauma patients. Intuitively, I also knew that Daniel was okay.

As soon as I walked into the house, the phone rang. The person on the other end was a nurse from the hospital in Forest Lake, and she told me exactly what I had intuitively known moments earlier. When she finished relating what had happened, I said calmly, "Yes, I already knew." I rushed to the small hospital in Forest Lake and was relieved to see that Daniel had not been injured. By the time I arrived, my wife was in an ambulance being transported to the other hospital.

When I picked up Daniel and hugged him, I realized there were small pieces of shattered windshield glass on his clothing—I even found pieces of glass in his diapers. As I drove to St. Paul, I wondered, *How did I know the details of the accident before I was told what happened? And, Was my foreknowledge the reason I remained calm at an emotionally disturbing time?*

Is this why precognition of traumatic events occurs?

Perhaps. In any case, I was grateful for the intuition, and I remember it well.

As I walked into the emergency room, my wife was being carted away for surgery. She looked relieved when she saw me with Daniel in my arms. I gently touched her forehead, and our eyes expressed relief and gratitude: things could have been a lot worse. Nurses and physicians had taken charge and were administering life-saving procedures as I left the emergency room hugging Daniel.

The story has a happy ending. My wife recovered completely.

Years later, while working on this book, I learned that my own clairvoyance and the many other extraordinary powers some individuals possess are not unusual. Next we will review some psi abilities we've discussed in earlier chapters and compare them to Patanjali's description of the *siddhis* to show that such "superhuman" powers were well known and documented in the *Yoga Sutras*.

Psi and Siddhis

In previous chapters we analyzed the nature of reality and who we are within the framework of extraordinary human abilities and recent findings in science. I suggested that these new findings were remarkably similar to the ancient texts of India. In this section we compare the *siddhis* described in the *Yoga Sutras* with the psi abilities discussed earlier to show that not only are they normal, but they are among our greatest attributes. Knowing about our extraordinary abilities informs us that we have the capacity to influence the realities of our lives. We are not at the mercy of a random life. Instead, we can create lives of peace, happiness, and abundance through the power of our infinite mind.

Following are excerpts from the *Yoga Sutras* paired with experiences we have examined through other lenses. I present the excerpts first.

Pada 1—Sutra 23
From trustful surrender to God, samadhi also comes.[32]

Pada 1—Sutra 25
Therein (in God) lies the seed of unsurpassed omni-science.[33]

These two sutras tell us that in surrender to God, one experiences *samadhi* or infinite joy, peace, love, and omniscience. The messages of these two sutras are similar to the NDEs of Eben Alexander and Anita Moorjani we examined in chapter 1. They both experienced bliss and a deeper knowing, and surrendered previously held notions about the nature of reality and life. Though their NDE experiences were like *samadhi*, they were not on the path of enlightenment when they went through them, which is not uncommon among NDE-ers.

Pada 3: Sutra 19 (I will abbreviate these notations in succeeding sutras; this one would be abbreviated 3:19.)

One can attain knowledge of others' minds.[34]
I gained knowledge of another's mind when I spontaneously knew what the football player at college was going to say before he spoke, as I described in chapter 3. Knowing about my ex-wife's car accident before the nurse informed me is also evidence of knowing the thoughts of others.

3:25—It is possible to obtain knowledge of subtle, concealed, and remote things.[35]
Pat Price's ability to describe the details of the Soviet atom bomb laboratory while sitting in a Faraday cage at SRI; dowser Harold McCoy finding the lost harp in Oakland, California, while he was in Arkansas; George McMullen's ability to describe archeological details; and my spontaneous ability to see a record player, speakers, light switch, and a picture of a woman on the cover of an LP are all examples of the human ability "to obtain knowledge of subtle, concealed, and remote things."

3:26–27—By performing samyama on the sun arises knowledge of the different realms in the universe.[36]
(By samyama) on the moon, knowledge of the solar system.[37]
Ingo Swann's ability to see rings around Jupiter while at SRI in Palo Alto, California, four months before NASA confirmed evidence of the rings in 1973 is an example of these two siddhis.

3:29—*(By samyama) on the naval plexus of the body comes knowledge of the arrangement of the body.*[38]
Joe Dispenza's healing of his shattered spine with the power of his thoughts, described in chapter 4, is an example of this *siddhi*.

3:38—*By loosening the cause of bondage, and by knowledge of the passageways of the mind, the mind can enter into the bodies of others.*[39]
Robert Rosenthal's experiment demonstrated that teachers' expectations that students will bloom intellectually raised students' IQ scores by 27 points—known as the Pygmalion effect. Also, Marva Collins's students did exceptionally well in their studies because she expected them to. Both of these examples from chapter 4 show that "mind can enter into bodies of others."

3:41—*By samyama on the relationship between the organ of hearing and the ether, divine hearing is attained.*[40]
Jay Greenberg, the young composer who began writing music at age two and had written five symphonies by age twelve, is an example of "divine hearing," discussed in chapter 3.

3:43—*The state of mind (projected) outside (of the body), which is not an imagined state, is called the great out-of-body (experience).*[41]
The out-of-body experience of Anita Moorjani

during her coma is an example of this sutra, from chapter 1.

3:44—*By samyama on the gross nature, essential nature, subtle nature, constitution, and purpose (of objects, one attains) mastery over the elements.*[42]
The telekinetic ability of Uri Geller we discussed in chapter 4 is an example of "mastery over the elements."

Siddhis for the Greater Good

Though the *siddhis* are widely seen as potential impediments to the goal of yoga in Indic traditions, Edwin Bryant states, "But not all *siddhis* are detriments to *samadhi;* after all, Patanjali (1:35) included supernormal sense experiences as suitable objects for the mind to concentrate on in order to achieve *samadhi.*"[43] And Pandit Rajmani Tigunait substantiates (1:35): "We have the ability to awaken the potential dormant in our cortex and eyes, putting their extraordinary powers at our command. This is true of all the senses and the capacities lying dormant within them." He adds, "Awakening these extraordinary powers anchors our mind, makes it flow peacefully, and accelerates our spiritual unfoldment."[44]

Furthermore, Bryant emphasizes, "Additionally Vyasa and Vacaspati Misra noted in 1:35 that upon experiencing some of the preliminary truths of *yoga,* the faith of the genuine yogi will thereby be reaffirmed and the commitment to proceed strengthened. A *yogi* sidetracked by them has clearly not mastered the *vairagya*, detachment, required as a preliminary to *yoga* (1:12)"[45]

Tigunait agrees and elaborates on the importance of experiencing extraordinary powers. "Spirituality is subtle. Because spirituality pertains to realities not perceptible to the senses, doubt is a substantial obstacle in spiritual practice. No matter how many scriptures we read or discourses we hear, without direct experience some degree of doubt regarding subtle spiritual matters invariably remains. But once we have direct experience of a reality beyond the ordinary domain of our senses, our belief in that particular aspect of reality—and in other aspects—unfolds naturally."[46]

The importance of *siddhis*, as Patanjali, yogis, and scholars explain it, is sound advice. My direct *siddhi* experiences not only triggered a keen interest in me to find answers to how and why they worked, but importantly, they helped me remain focused on my inward journey of meditation. I learned that Patanjali's advice neither to be seduced by the *siddhis* nor to shun them was not contradictory.

There are compelling reasons for not getting sidetracked by the *siddhis*, and though they remain dormant for most, it still behooves us to ask why we shouldn't use our inherent superhuman gifts for living radically healthier, happier, and more purposeful lives. What should people who are born with superhuman powers do with them? Should shamans and healers not use their *siddhis*? How should they understand and use their gifts?

Since Patanjali asserts that *siddhis* are not always impediments to the goal of yoga, attainment of some physic abilities can and should be interpreted without doubt or confusion. With this clarity, our psychic abilities to heal ourselves and others can be seen as gifts meant to be used. Using *siddhis* to gain knowledge about ourselves and the universe would be considered prudent. And not using our superhuman powers to create greater love, peace, and harmony in our inner and outer worlds would be an abdication of responsibility.

Living an Abundant Life

In the mid-nineteenth and early twentieth century, writers like James Allen, Joseph Murphy, and Neville Goddard, among others, brought forth the idea that the *infinite power of mind creates the outer world*. According to them, it is possible to attain whatsoever we desire through the power of our thoughts when we wish with clarity, belief, gratitude, and a joyous heart. We will briefly review these writers and their work for two reasons: their insights mirror the perennial philosophy, and their work corresponds to key concepts of quantum physics.

James Allen was born in Leicester, England, in 1864. His 1903 work *As a Man Thinketh* earned him worldwide fame as a prophet of inspirational thinking. In it, he begins by stating his philosophy: "Mind is the Master power that moulds and makes, And Man is Mind, and everyone he takes. The tool of Thought, and, shaping what he wills, Brings forth a thousand joys, a thousand ills: He thinks in secret, and it comes to pass: Environment is but his looking glass."[47]

In these poetic words, Allen captures the essence of his

thesis: the thoughts, feelings, and emotions of our inner world mirror the observations and experiences of our outer world.

The popularity of his book was due not only to his being an excellent writer, but also because his philosophy reflected the words of the Bible. He wrote, "The aphorism, 'As a man thinketh in his heart so is he,' not only embraces the whole of a man's being, but is so comprehensive as to reach out to every condition and circumstance of his life. A man is literally *what he thinks,* his character being the complete sum of all his thoughts."[48]

It is interesting that he had come to know what quantum theory would suggest a few years later. His thinking influenced a who's who of self-help writers, including Napoleon Hill, Dale Carnegie, and Norman Vincent Peale. Though his books on creating an abundant life were especially popular among Christians, his message was well received by many others throughout the world and is still popular.

In the early 1940s, Joseph Murphy wrote *The Power of Your Subconscious Mind,* which became one of the most powerful self-help books of the mid-twentieth century. Similar to James Allen's message, Murphy also claimed that thoughts create the outer world. By contextualizing his philosophy and practical guidance in secular terms, such as "*conscious* and *subconscious mind*," he reached a wider audience.

Murphy wrote, "Whatever thoughts, beliefs, opinions, theories, or dogmas you write, engrave, or impress on your subconscious mind, you will experience them as the objective manifestation of circumstances, conditions, and events. What you write on the inside, you will experience on the outside. You have two sides to your life, objective and subjective, visible and invisible, thought and its manifestation."[49] His philosophy, though primarily expressed in secular terms, was also framed in the words of the Bible: "Ask, and it shall be given you: seek, and ye shall find; knock, and it shall be opened on to you." (Matthew 7:7)

Murphy's emphasis was on how to access the subconscious mind. His instructions were straightforward. "You must ask *believing*, if you are to receive." By this he meant any desired wish should be imagined as if it has already been fulfilled. Furthermore, he added, when the infinite subconscious mind accepts an idea, its manifestation is inevitable.[50]

Murphy's ideas had earlier been expressed by Neville Goddard. Mitch Horowitz, in the introduction to *The Neville Reader* states, "Neville was among the last century's most articulate and charismatic purveyors of the philosophy generally called New Age."[51] In the late '30s he traveled extensively to share his philosophical ideas based on the scriptures to enthusiastic audiences in churches and packed auditoriums all over America. Though his philosophy was rooted in the scriptures, his interpretations were metaphorical. He believed that the physical world was literally a reflection of our thoughts, and therefore it was possible for us to create any reality we truly desired.

As a young man he had worked as an actor in New York, but was not successful at it. Later he wrote, "After 12 years in America I was a failure in my own eyes."[52]

Brokenhearted and with empty pockets, Goddard

longed to visit Barbados, where he had grown up, for Christmas. His friend Abdullah, with whom he studied Hebrew, scripture, and Kabbalah for five years, told him that if he imagined with feeling being with his family at Christmas, he would.[53]

On a December morning before the last ship was to depart for Barbados, he received a check for $50 from his long-out-of-touch brother, making his trip possible. This remarkable experience eventually became the hallmark of his philosophy: "It is not what you want to attract, you attract what you believe to be true."[54] Goddard's belief in manifesting what we desire by imagining it to be true in the present has scientific merit. As we discussed earlier, thoughts not only bring into existence subatomic particles, but they also create the material world of sticks and stones.

Goddard's emphasis on feeling the future in the present was similar to Allen's and Murphy's. His advice to harness what we desire emphasized that one must hold a thought with concentration while in a relaxed frame of mind. In his lectures he explained that the best times to imagine that what we desire has already manifested are just before going to sleep and just before waking up, when we are relaxed and drowsy. And he suggested we should repeat this desire in our mind until it's attained. Goddard was well aware of the necessity to wish for an outcome in a relaxed manner, and he encouraged its practice, just as Allen and Murphy had done. Though all three men wrote about the importance of meditation, their emphasis lacks the significance given to it in the *Bhagavad-Gita* and the *Yoga Sutras*.

Goddard's assertion that thoughts create the physical world was a remarkable idea in his day. Though he was correct, questions remain: Why does it not work for everyone? And why doesn't it work always?

There are no definitive answers to these questions. We live in a magical world beyond our sense-based conceptions of reality. Therefore, it is best to experience life in the inner world and build a bridge to the outer world. All of life is but a dream, so we must dream dreams in which the world is good and beautiful in every way. It is entirely possible to manifest such a world! The way to attain whatever we desire is described in Allen's, Murphy's, and Goddard's inspirational and informative books. I recommend reading any one of them as a follow-up to this book.

When we learn about violence and bloodshed, wars and famine, racism and injustice, or the devastating effects of climate change, we may become disheartened and dejected, and feel helpless. We may think, *What can one person do to change the world?* When we struggle with poor health or unfortunate circumstances, we may succumb to feelings of hopelessness and despair. But we may not realize that each of us, with the power of our thoughts, can create whatever we desire including a better world.

Some years back, in the midst of anguish and helplessness about the problems of the world, and not knowing what to do, I decided to go inward, and I discovered that the outer world is a reflection of my thoughts and emotions. This gave me a real sense of hope! The well-being I felt as a result of meditation, study, and reflection had an impact on the outer world. The impact may have been small, but it felt meaningful. *What if*, I thought, *greater numbers of people did the same?*

Finding an Artist

I spent hours and hours thinking about an illustration for this book's cover. What should it depict? Where would I find an artist to create it? And how much would it cost? I searched stock illustrations on the Internet for ideas and found a few that were similar to what I had in mind, but I needed an artist to draw what I felt would be more suitable for this book.

I wasn't interested in contacting a professional illustrator, because I assumed the fee would be more than I had in my budget. Thinking it might be a more affordable option, I asked a couple of friends if they knew of anyone who worked as an artist to supplement their income. My friend Mark knew of such a person, his neighbor—I was in luck! But the neighbor wasn't interested because he was too busy.

With regret, I decided to forgo an illustration for the cover, but kept alive the possibility of finding an artist, and continued with life in the spirit of "Thy will be done."

On the Saturday after I learned Mark's neighbor wasn't interested, I planned on playing tennis with Mark. His wife was using their only car, so I volunteered to give him a ride to the club where we play, even though he lives about thirty miles from my place. I got to his house fifteen minutes early, and he wasn't ready yet, so I decided to take a walk in an adjacent park, a beautiful vantage point that overlooks downtown St. Paul and the Mississippi River.

As I entered the park, I saw an artist about thirty yards away, painting the scene below. Immediately, I felt he would be the one who could help me with the illustration. With excitement and anticipation, I approached him and saw the painting he was working on. I complimented him on his beautiful depiction of the city beside the river, and as we talked, I learned he was a busy and successful artist.

I asked him if he would be interested in creating an illustration for my book's cover—and he was! Then I told him I was on a limited budget, and was sure I wouldn't be able to pay him what he was worth, but he didn't seem concerned. In fact he seemed eager to work on my project and handed me his business card. The artist who created the illustration for this book's cover is Tom Stewart—the artist I met in the park.

This experience once again reminded me that thoughts create reality. As Allen, Murphy, and Goddard might have explained, my thoughts and desires had nudged the universe to produce exactly what I needed. The universe is indeed like a great thought, as physicists have proclaimed, and our thoughts function in the domain of a universal mind. *Atman and Brahman are one, We are made in His Image*, and *The Kingdom of Heaven resides within us*. This is how and why our thoughts create the external world.

Next time you experience a coincidence, reflect upon it and be grateful. Know that your imagination and desire created it. There is no doubt you can manifest your desires, but also know, as Matthieu Ricard cautions, "The universe is not a mail order catalog." But it is friendly, and mirrors our deeply held convictions and desires. So dream big and consider following the three pillars of yoga; knowledge, experience, and practice. *Knowledge* goes in search of the big questions. *Experience* refers to ordinary and nonordinary (such as clairvoyance, divine intervention, and *samadhi*) incidences in our lives, and *practice* means meditation, silence, and contemplation (yoga).

A Parting Word

The insights and understandings I have illuminated throughout this book—found in ancient wisdom traditions and validated by new scientific findings—are not meant to be definitive. Rather, they are intended to inspire reflection. I hope this book proves a useful guide for you as you pursue greater knowledge, learn from your experiences, and develop your own practice. I wish you peace, love, and happiness throughout your journey.

Epilogue:
Imaginal Activism—
A Promise to the River

The youngest prophet of the climate crisis, Greta Thunberg, addressed the United Nations Climate Action Summit on December 15, 2018: a speech that resonated with children and many adults all over the world. An untold number of viewers saw her on television and 1.5 million others heard her speech online. In the summer of 2019 she arrived by sailboat on the shores of New York City to spill her heart to adults: *"Yet you all come to me for hope? How dare you?"*

Thunberg's sentiments stirred much attention, but it was her palpable love and genuine grief that touched the collective heart. Spilling out into the streets, heartbroken, passionate young people the world over are building momentum for climate awareness. What makes this movement so powerful, and strange, is that it is through the mouths of babes that we are waking up to the unthinkable reality of the end of life on Earth as we know it—we are living in the shadow of death.

* * *

The river at the Coon Rapids Dam and County Park was much higher than normal and flowing fast. It had rained for several days prior to this crisp sunny October afternoon, causing the rapid flow and near flood-stage conditions. The walking trail beside the river is full of trees, shrubs, and wildflowers, which, oddly, were blooming in October. The erratic weather—a cooler than normal summer and a warmer than normal fall—was the reason for these unusual blooms. Though they looked lovely, I was concerned. As I walked along the river basking in the sunshine, I saw caterpillars that are normally present in the spring; it was odd to see them just before winter. *Surely they will perish*, I thought, *as soon as the first snowfall*. Snow was forecast for the following week. My dismay and concern for the caterpillars reminded me that the use of pesticides has decimated insect populations, resulting in the reduction of birds by the billions over the past few decades. However, much to my relief, a few days later, I learned that some of the orange and black woolly bear caterpillars would survive the winter.

Along this path adjacent to the banks of the Mississippi River, I saw, as I have on many other occasions, white foam on its surface, similar to soap bubbles floating on bathwater. The Army Corps of Engineers, which is responsible for reporting on the environmental health of the river, maintains that the white foam is the result of the decay of organic materials such as sewage, plants, and grass clippings. But I couldn't help but think about the tons of fertilizer and pesticide runoff from farms upstream, which have caused dead zones in the Gulf of Mexico in an area approximately the size of New Jersey: eighty-five hundred to nine thousand square miles. The runoff contains nitrogen and phosphorus, reducing oxygen levels and killing fish and marine life. The "mighty" Mississippi is causing massive algal growth, and there is no sign this will end, or even slow down.

Thoughts of Rachel Carson's *Silent Spring,* the decimation of insects, the significant reduction of birds, and the ever-increasing problem of dead zones in the ocean made me stop and look at the river with deep concern. In my heart I promised the river that I would help restore it to how it used to be. Not how it was just a few decades ago, but in its pure state five hundred years ago, when its water was so clean that I could have cupped my hands, reached in, and taken a drink.

What caused me to shift my normal outlook of concern and worry about the river, the environment in general, and the devastating effects of climate change? Why did I make this promise, which was also a promise to the children of the world so they can inherit a livable and healthy planet? It seemed outlandish from a rational and materialist perspective, but it was perfectly normal in the context of infinite mind—and making this promise felt good. Silently in my mind, I spoke to Greta Thunberg and shared my pledge to the river with her.

I had made the promise, but how can we begin to

approach the unthinkable? How are we to proceed? First, we must not pretend that we can solve the climate emergency with the urgent tools of the old paradigm. Instead we must realize that we are in oneness with the universe and that we possess inner gifts of imagination that will enable us to recreate the world as it once was. A world in which the rivers and lakes are pure and clean. A lush, green world in which millions of insects, birds, animals, and fish thrive. A world of abundance, diversity, and harmony.

Accepting deeply and without illusion that human activities have already led to irreversible, unthinkable consequences is the first step toward an activism without false hope. Next we must believe we can recreate our world as it once was using the innate and wondrous gift of our infinite mind. This is not an either-or proposition. Implementing strategies to eliminate the use of fossil fuels; converting to sustainable and renewable energy sources; and creating a new economic model as Bill McKibben, Lester Brown, Paul Hawken, and others have proposed must be done urgently. And we must pay attention to Greta Thunberg's plea for immediate action, as she expressed in her address to members of the British Parliament on April 23, 2019.

> Around the year 2030, ten years, two hundred fifty-two days, and ten hours away from now, we will be in a position where we set off an irreversible chain reaction beyond human control that will most likely lead to an end of our civilization as we know it. That is unless in that time, permanent and unprecedented changes in all aspects of society have taken place.

Did you hear what I just said? Is my English okay?
Is the microphone on? Because I'm beginning to
wonder.

Though young Thunberg's message has been heard
around the world—not only by children but also by
adults—in a way that it has not been heard before, there
exists a feeling in the hearts and minds of many people
that whatever we actually do to avert the effects of climate
change will be too little too late. As I witness world leaders'
current lack of action and policies in addressing humanity's
existential crisis, I am alarmed, and I empathize with the
scientists and concerned citizens of the world. So, what
are we to do?

Throughout this book we have discussed the power
of our infinite mind and how we create the outer world
according to what we think, feel, and nurture in our inner
world. With this understanding, I would like to leave you
with this radically simple suggestion: *imaginal activism.*
How do you do imaginal activism? All you have to do is
meditate with intent, belief, and a joyous heart, *imagining*
a "new Earth" in the likeness of how it was five hundred
years ago. (An imaginal activism meditation you can use
is included in appendix B.) Repetition is critical, so every
time you meditate, imagine pristine and perfect conditions
for all of nature and the Earth.

There is no downside to this practice, even if we don't
attain exactly what we wish. But what if we did? Our collective thoughts would create the outer world to reflect
what we assume, and we would attain a livable world for

our children, grandchildren, and everyone else for a long, long, time to come. Together, with the power of our minds we can keep my promise to the river—and to all of God's children.

Miracles happen! God creates! Our intentions with belief in a state of quiescence will work because "I am That"—"So Hum."

Appendix A:
Nature of Reality

Emergent Thinking	Conventional Thinking
Consciousness creates the brain.	Brain creates consciousness.
Life is infinite.	Life is finite.
There is no past, present, or future. Time and space coexist as space-time. Time is an illusion. The present is eternal.	Time is a linear progression of past, present, and future. Time exists as an independent phenomenon.
Universal mind is omnipresent, omniscient, interconnected to all minds, and is their source. It is nonlocal, infinite, and ever changing.	Each human mind is independent. All thoughts, ideas, and insights are created in the brain.

Emergent Thinking	Conventional Thinking
The brain is a filtering mechanism. It decodes the vibrational energy present in the world through the sense organs, and creates the illusions of sight, sound, etc., which we consider as reality.	External objects and all other activities of the world have independent existence that is observed through the sense organs.
Infinite possibilities exist in the universe. Our thoughts, words, and actions create the realities we experience.	There are no possibilities "out there." Life and our experiences happen by chance. Our thoughts, words, and deeds have some influence on the realities we experience.
The power of the infinite mind exists in all humans. However, these powers remain dormant for most of us, though they can be accessed through meditation, yoga, and prayer.	There is no such thing as infinite mind. Knowledge, ideas, and creativity require the hard work and effort of our independent brains.

Emergent Thinking	Conventional Thinking
Mind-to-mind communication is possible and fundamental. Neither energy nor time is expended in this form of communication.	Communication is only possible through speech, the written word, or images (such as through art and dance).
DNA has energetic vibrational stability. It can be reshaped by our dominant thoughts, feelings, and beliefs.	DNA is a solid blueprint. It affects all aspects of our life. It is not dependent on our thoughts, feelings, or beliefs. Some aspects of DNA lie dormant and can be turned on or off based on external stimuli (i.e., diet, exercise, etc.)

Emergent Thinking	Conventional Thinking
The scientific method is limited in scope. It is useful for inquiry into phenomena in the local domain of the 3-D world. However, since the worlds of locality (matter) and nonlocality (thought) are intertwined, it's not possible to delve deeper into the nature of reality using this method.	The scientific method of research is reliable. It has contributed significantly to the knowledge and understanding of our world. This method of inquiry, based on the philosophy of scientific materialism, will continue to add new knowledge and understandings about the nature of reality.
Consciousness is the source of the universe. Everything in it is vibrational energy, some of which appears as matter. The "Big Bang" that seems to have created the universe is one among countless others.	The universe consists only of matter and energy, and originated with the "Big Bang." It isn't possible to know what existed before the "Big Bang."

Emergent Thinking	Conventional Thinking
The binary logic of reason, language, or how we think is incomplete. Tetralemma logic is comprehensive and explains the observations of quantum theory, as well as of the macro world.	Binary logic is complete. It cannot explain the "hard problem" of consciousness or the "weird" nature of quantum theory. However, it will be able to do so in the future.
The universe was not created by chance. Elements of design, coherence, and purpose are evident in its nature.	The universe was created by chance. Elements of design and coherence have also happened by chance. There is no evidence of purpose in the universe.

Appendix B:
Imaginal Meditation

I njustice, war, famine, genocide, extortion, and other forms of abuse are among the wrongs that exist in the world. Ending them demands our attention, resolve, and concerted action. The one that looms largest, beyond all others, is climate change. This *existential crisis* requires an all-hands-on-deck approach because the time to address it is running out. In the epilogue I suggested that in addition to the strategies currently being used to address the climate emergency, another approach is needed: using the power of our infinite mind by entering the inner domain.

Imaginal Activism

Our infinite mind is the most powerful force we can bring to bear on reversing the effects of climate change. We can access it during the quiescence reached in meditation and by imagining the state desired. Using imaginal activism to avert climate change simply means imagining a healthy and sustainable world. We can begin by entering into a future in which the natural world is as pristine as it was five hundred years ago. When we are present in this future with appreciation, gladness, gratitude, and wonder, we fall in love with it. By entering this pure world in the privacy of our hearts and minds, we realize there is no downside to it. And in experiencing this enchanted and perfect inner world, we realize it is the source of what we experience in the outer world. We don't need to know how it works. We don't have to construct or change anything deliberately. Our imagination creates the outer world much as Pygmalion's falling in love with his statue brought it to life.

Let's enter this magically perfect world now.

An Enchanted Forest and River

Find a quiet place to meditate. Sit in the lotus position or on a chair. Relax. Calmly breathe in and out.

When you reach a state of quiescence, imagine being in an enchanted forest on the shore of a river. It is peaceful here. The river is calm, and the forest is serene and beautiful. You feel blessed to be here and you continue to breathe with ease. You feel like moving closer to the river, and you do. You are close enough to touch it, and you do. It feels soft, wet, and cool. You are delighted by this feeling and you smile. You run your fingers gently through the surface of the water, which is clean enough to drink. You are at peace, and you want time to stop. Sun filters through the leaves of the trees lining the riverbank, casting shadows intermingled with sunshine.

You wonder how it would feel to step into the water.

You can't resist the temptation. The river's beauty and calm intertwine with yours, and you take a few steps into the water. The soft sand under your feet and the water lapping your ankles calm you even more. You stand here, continuing to breathe comfortably, aware of yourself and everything around you. You are in awe of the pristine nature of the river and the forest. Everything seems abundant and alive.

Stay here for as long as you like. Then, after you have felt the serenity and peace of the river, come back to its shore. There is a warm, clean towel waiting for you. Feel the softness of the towel as you dry your feet and ankles. Don't rush. Be present and feel every sensation. You are comfortable, warm, content, and at peace. As you look around, you see that the trees are lush and green. They look healthy, strong, and sturdy. You walk up to one near you and put your hand on its bark. It feels good to touch the hard wood. It gives you a feeling of sturdiness and strength.

You feel that everything is as it should be. In this relaxed, comfortable, peaceful, and secure moment, you hear birdsong. The sound of many different species of birds in the surrounding trees makes the forest come alive. Now you notice butterflies fluttering around you. One of them lands on a flower and then flies to another one.

You feel that you have been in this enchanted forest before. You feel you belong here—it is perfect in every way. As you stroll through the woods, you feel that the entire Earth must be full of such beautiful places. In your transcendent state of bliss, your gaze turns to another butterfly as it lands on your shoulder. You take a deep breath to appreciate the presence of this tiny creature.

You want it stay on your shoulder, and it does. You find an old log on the ground and decide to sit down. All is well with you and the world. You feel connected to everything around you. It is the kind of connectedness you felt as a child when your mother held you in her arms, but even better. It is a feeling where you desire nothing because you have everything. You are filled with peace, joy, and deep gratitude. You feel what it is like to be truly you. And having created and experienced this enchanted and pristine world, you realize with complete and full conviction that the outer world will reflect what you experienced in your inner world.

Now you are ready to come back to the external world. As you open your eyes, everything still looks the same, but you know in your heart that you have nudged the universe. You and hundreds and thousands or millions of people imagining as you have done can indeed transform the world to the one you imagined. You have done all this without knowing how it actually works. You have done it in the comfort and safety of your own home. You have done it knowing that the infinite mind creates the reality you have dreamed and expect. And, you have done it with attached-detachment.

Notes

Chapter 1

1. Patricia Hampl, *I Could Tell You Stories: Sojourns in the Land of Memory* (New York: W. W. Norton and Company, 1999).

Chapter 2

1. Stephanie Pappas, "Hitchcockian Crows Spread the Word About Unkind Humans." *Live Science*, June 28, 2011. https://www.livescience.com/14819-crows-learn-dangerous-faces.html

2. Robert Roy Britt, "No Birdbrain, Parrot Grasps Concept of Zero." Live Science, July 8, 2005. http://www.livescience.com/3907-birdbrain-parrot-grasps-concept.html

3. "Inside Animal Minds: Who's the Smartest?" *NOVA*, April 23, 2014. Segments of this PBS program, in which crows and other birds are shown solving complex puzzles, can be seen on YouTube.

4. University Of New Hampshire, "Researcher Uncovering Mysteries of Memory by Studying Clever Bird." *ScienceDaily*, October 12, 2006. www.sciencedaily.com/releases/2006/10/061012094818.htm

5. Stephan Harrod Buhner, *Plant Intelligence and the Imaginal Realm* (Rochester, VT: Bear and Company, 2014).

6. University of Melbourne, "Pigeons know the difference between good and bad art," August 24, 2016. https://blogs.unimelb.edu. au/sciencecommunication/2016/08/24/pigeons-know-the-difference-between-good-and-bad-art/

7. Maddie Stone, "New Brain Scans Show How Dolphins Use Sound to See." gizmodo.com, July 8, 2015. https://gizmodo. com/dolphin-brain-scans-show-how-vision-and-hearing-are-con-1716228432

8. Suzanne W. Simard, "Net transfer of carbon between ectomy-corrhizal tree species in the field." *Nature* 388 (August 7, 1997), 579–82.

9. Stephan Harrod Buhner, *The Lost Language of Plants* (White River Junction, VT: Chelsea Green, 2002); Peter Wohlleben, *The Hidden Life of Trees* (Vancouver, BC: Greystone, 2015).

10. Ibid. (Both Buhner and Wohllenben, in their respective books.)

11. Buhner, *The Lost Language of Plants.*

12. Evelyn Keller, *A Feeling for the Organism: The Life and Work of Barbara McClintock* (New York: Holt, 1983), 140.

13. Ibid., 117

14. Ibid., 69.

15. Ibid., 117.

16. Ibid., 125.

17. Cleve Backster, "Evidence of a Primary Perception in Plant Life," *International Journal of Parapsychology*, vol. 10, no. 4 (Winter 1968), 329.

18. Cleve Backster, *Primary Perception: Biocommunication with Plants, Living Foods, and Human Cells* (White Rose Millennium Press, 2003).

19. Ibid.

20. Wikipedia, "Oldest trees." https://en.m.wikipedia.org

21. Stefano Mancuso, *The Revolutionary Genius of Plants: A New Understanding of Plant Intelligence and Behavior* (New York: Simon and Schuster, 2018), 5.

22. Ralph Waldo Emerson, *Essays*, 1841.

23. Dawson Church, *Mind to Matter* (Carlsbad, CA: Hay House, 2018).

24. Ibid.

25. Ibid.

26. Masaru Emoto, *The Hidden Messages in Water* (Hillsboro, OR: Beyond Words, 2001). [AU: please add page number]

27. Bernd Kröplin and Regine Henschel, *Water and Its Memory: New Astonishing Insights in Water Research* (Stuttgart: Gutes-Buch Verlag; first English edition 2017).

Chapter 3

1. Plato, The Allegory of the Cave, *The Republic*.

2. Russell Targ, *The Reality of ESP: A Physicist's Proof of Psychic Abilities* (Wheaton, IL: Quest, 2012).

3. Ibid.

4. Ibid, 24.

5. Ibid.

6. Ibid., 27.

7. Ibid.

8. Ibid.

9. Ibid., 31–32.

10. Goddard Space Flight Center, "How Did Jupiter's Rings Form?" https://www.nasa.gov/centers/goddard/multimedia/largest/rings.html

11. Russell Targ and Jane Katra, *Miracles of Mind: Exploring Nonlocal Consciousness and Spiritual Healing* (Novato, CA: New World Library, 1998), 40.

12. Targ, *The Reality of ESP*, 51.

13. Ibid., 53.

14. Ibid.

15. Ibid., 57.

16. Ibid.

17. Russell Targ and J. J. Hurtak, *The End of Suffering: Fearless Living in Troubled Times* (Charlottesville, VA: Hampton Roads, 2006).

18. Targ and Katra, *Miracles of Mind*.

19. Elizabeth Lloyd Mayer, *Extraordinary Knowing: Science Skepticism, and the Inexplicable Powers of the Human Mind* (New York: Bantam, 2007).

20. Ibid., 2–3.

21. Ibid.

22. Ibid., 5.

23. Ibid.

24. Thomas Kuhn, *The Structure of Scientific Revolutions* (Chicago: University of Chicago Press, 1962).

25. Mayer, *Extraordinary Knowing*, 8.

26. Stephan A. Schwartz website: http://www.stephanaschwartz.com

27. Stephan A. Schwartz, *The Secret Vaults of Time: Psychic Archeology and the Quest for Man's Beginnings* (New York: Grosset and Dunlap, 1978), 108.

28. Ibid.

29. Ibid.

30. Ibid., 136–37.

31. Ibid., 138.

32. Ibid., 149.

33. Ibid.

34. Stephan A. Schwartz, *Opening to the Infinite* (Langley, WA: Nemoseen Media, 2007).

35. Ibid.

36. Ibid.

37. Ibid., 199.

38. Ibid.

39. Wikipedia, "Jay Greenberg (composer)." https://en.wikipedia.org/wiki/Jay_Greenberg_(composer)

40. CBS, *60 Minutes,* November 28, 2004.

41. Ibid.

42. Ibid.

43. Nancy L. Segal, *Born Together—Reared Apart: The Landmark Minnesota Twin Study* (Cambridge, MA: Harvard University Press, 2012).

44. Ibid.

45. Ibid., 28.

46. Ibid., 27.

47. Ibid.

48. Darold A. Treffert, "The Savant Syndrome: An Extraordinary Condition. A synopsis: past, present, future," *Philosophical Transactions of the Royal Society B: Biological Sciences,* Vol. 364, no. 1522, 1351–57 (May 27, 2009). https://www.ncbi.nlm.nih.gov/pmc/articles/PMC2677584

49. Darold A. Treffert, *Extraordinary People: Understanding Savant Syndrome* (New York: Ballantine, 1989).

50. Treffert, "The Savant Syndrome."

51. Treffert, *Extraordinary People.*

52. CBS, *60 Minutes,* December 19, 2010.

53. Deepak Chopra and Rudolph Tanzi, *Super Brain: Unleashing the Explosive Power of Your Mind to Maximize Health, Happiness, and Spiritual Well-Being* (New York: Three Rivers, 2012), 104.

54. Ibid.

Chapter 4

1. Sir James Jeans, *The Mysterious Universe* (Whitefish, MT: Kessinger, 2010), 137.

2. Wikipedia, "Pygmalion (mythology)." https://en.wikipedia.org/wiki/Pygmalion_(mythology)

3. Katherine Ellison, "Being Honest About The Pygmalion Effect," *Discover,* December 2015. http://discovermagazine.com/2015/dec/14-great-expectations

4. Ibid.

5. Robert Rosenthal and Lenore Jacobson, *Pygmalion in the Classroom: Teacher Expectation and Pupils' Intellectual Development* (New York: Holt, Rinehart and Winston,1968).

5. Ellison, "Being Honest About The Pygmalion Effect."

7. Marva Collins and Civia Tamarkin, *Marva Collins' Way* (New York: Tarcher, 1982), 21.

8. Ibid., 22

9. Ellen Langer, *Counterclockwise: Mindful Health and the Power of Possibility* (New York: Ballantine, 2009).

10. Fabrizio Benedetti, "Placebo Effects: From the Neurological Paradigm to Translational Implications," *Neuron,* Vol. 84, issue

3 (November 25, 2014), 623–37. https://doi.org/10.1016/j.
neuron.2014.10.023

11. Harvard Men's Health Watch, "The Power of the Placebo
Effect," Harvard Health Publishing, May 2017. https://www.
health.harvard.edu/mental-health/the-power-of-the-placebo-
effect

12. Benedetti, "Placebo Effects."

13. Ibid.

14. Joe Dispenza, *Evolve Your Brain: The Science of Changing Your
Mind* (Deerfield Beach, FL: Health Communications, 2007), 18.

15. Dawson Church, *Mind to Matter: The Astonishing Science of
How Your Brain Creates Material Reality* (Carlsbad, CA: Hay
House, 2018)

16. Ibid.

17. Ibid., 38–39.

18. Ibid., 43.

19. Ibid.

20. Edgar Mitchell, *The Way of the Explorer: An Apollo Astronaut's
Journey Through the Material and Mystical Worlds* (New York:
Putnam, 1996), 82.

21. Ibid., 83.

22. Ibid., 87.

23. Ibid., 94.

24. Ibid., 88.

Chapter 5

1. Manjit Kumar, *Quantum: Einstein, Bohr, and the Great Debate
about the Nature of Reality* (New York: Norton, 2008).

2. Ibid., 262.

3. Wikipedia, "Complementarity (physics)." https://en.m.wikipedia.org/wiki/Complementarity_(physics)

4. Jeans, *The Mysterious Universe*, 137.

5. John Archibald Wheeler, "Niels Bohr and Nuclear Physics," *Physics Today*, Vol. 16, No. 10 (October 1, 1963), 36.

6. Dalai Lama, *The Universe in a Single Atom: The Convergence of Science and Spirituality* (New York: Morgan Roads Books, 2005), 46.

7. Ibid., 47.

8. Arthur Stanley Eddington, *The Nature of the Physical World* (New York: MacMillan, 1930), 291.

9. Dalai Lama, *The Universe in a Single Atom*, 50.

10. Bernardo Kastrup, *The Idea of the World: A Multi-Disciplinary Argument for the Mental Nature of Reality* (Alresford, Hants, UK: Iff Books, 2019), 6.

11. Ervin Laszlo, *What Is Reality? The New Map of Cosmos and Consciousness* (New York: SelectBooks, 2016).

12. Jill Bolte Taylor, *My Stroke of Insight: A Brain Scientist's Personal Journey* (New York: Plume, 2006), 19.

13. Joseph Bennington-Castro, "Want Ultraviolet Vision? You're Going to Need Smaller Eyes," Gizmodo.com, November 22, 2013, https://io9.gizmodo.com/want-ultraviolet-vision-youre-going-to-need-smaller-ey-1468759573

14. Laszlo, *What Is Reality?*

15. Taylor, *My Stroke of Insight,* 19–20.

16. "How Well Do Dogs and Other Animals Hear?" Louisiana State University. https://www.lsu.edu/deafness/HearingRange.html

17. Taylor, *My Stroke of Insight,* 20.

18. Ibid.

19. Eben Alexander, *Living in a Mindful Universe: A Neurosurgeon's Journey into the Heart of Consciousness* (New York: Rodale, 2017).

20. Ibid., 47.

21. Ibid., 48.

22. Ibid.

23. Rupert Sheldrake, *Morphic Resonance: The Nature of Formative Causation* (Rochester, VT: Park Street, 2009), 212–13.

24. Rupert Sheldrake, *The Sense of Being Stared At: And Other Aspects of the Extended Mind* (Rochester, VT: Park Street, 2013).

25. Alexander, *Living in a Mindful Universe*, 63.

26. From a letter written to the widow of a recently deceased friend. Reprinted in Carolyn Kormann, "What Does a Minute Feel Like"? Culture Desk, *The New Yorker*, August 7, 2014. https://www.newyorker.com/culture/culture-desk/minute-feel-like#targetText=That%20means%20nothing.,only%20a%20stubbornly%20persistent%20illusion.%E2%80%9D

27. Dalai Lama, *The Universe in a Single Atom*, 60.

28. Wikipedia, "Hafele-Keating experiment." https://en.wikipedia.org/wiki/Hafele%E2%80%93Keating_experiment

29. Taylor, *My Stroke of Insight*.

30. Ibid., 13.

31. Ibid., 18.

32. Dalai Lama, *The Universe in a Single Atom*, 48.

33. Targ and Hurtak, *The End of Suffering*, 46.

34. Ibid., 45.

35. Mitchell, *The Way of the Explorer*, 57–59.

36. This version of the Manifesto for a Post Materialist Science is a summarized statement from Laszlo's *What Is Reality?*, 41.

37. Ervin Laszlo, *The Intelligence of the Cosmos: Why Are We Here? New Answers from the Frontiers of Science* (Rochester, VT: Inner Traditions, 2019).

38. Ibid., 33.

39. Ibid., 4.

40. Fred Hoyle, *The Intelligent Universe: A New View of Creation and Evolution* (New York: Holt, 1988). https://www.goodreads.com/author/quotes/199992.Fred_Hoyle

41. Laszlo, *The Intelligence of the Cosmos.*

42. Ibid., 36.

43. Ibid.

44. Targ and Katra, *Miracles of Mind*, xv.

45. Laszlo and Currivan, *CosMos,* 5.

46. Ibid.

47. Ibid.

48. Laszlo and Curriva, *CosMos.*

49. Brian Greene, *The Hidden Reality: Parallel Universes and the Deep Laws of the Cosmos* (New York: Random House, 2011), viii.

50. NASA Science Mission Directorate, "Sweet Meteorites," December 20, 2001. https://science.nasa.gov/science-news/science-at-nasa/2001/ast20dec_1

51. Laszlo, *The Intelligence of the Cosmos*, 33.

52. Ibid.

53. University of Colorado at Boulder, "New Study hints at spontaneous appearance of primordial DNA,"

Science Daily, April 7, 2015. https://www.sciencedaily.com/releases/2015/04/150407095635.htm

54. Bruce Lipton, *The Biology of Belief: Unleashing the Power of Consciousness, Matter, and Miracles* (Carlsbad, CA: Hay House, 2005).

55. "Number of genes in human genome lower than previously estimated," *MIT News,* October 29, 2004. http://news.mit.edu/2004/humangenome

56. Lipton, *The Biology of Belief.*

57. David Baltimore, "Our Genome Unveiled," *Nature* 409 (February 15, 2001), 814–16. https://www.nature.com/articles/35057267

58. Lipton, *The Biology of Belief,* 188.

59. Alan H. Batten, "A Most Rare Vision: Eddington's Thinking on the Relation Between Science and Religion," *Quarterly Journal of the Royal Astronomical Society,* Vol. 35, No. 3 (September 1994), 249.

60. "Look for Truth—No Matter Where It Takes You": interview with F. David Peat on David Bohm, Krishnamurti, and Himself in *What Is Enlightenment?* Vol.6, No. 1 (Spring/Summer 1997), 18. http://www.fdavidpeat.com/interviews/wie.htm

Chapter 6

1. Aldous Huxley, *The Perennial Philosophy* (New York: Harper-Collins, 1944), viii.

2. Huxley, *The Perennial Philosophy.*

3. Ibid.

4. Edward Waldo Emerson and Waldo Emerson Forbes, eds., *Journals of Ralph Waldo Emerson,* (Boston: Houghton Mifflin, 1909), 7:11.

5. Henry David Thoreau, *Walden* (Seattle: Amazon Classics), 156.

6. *The Bhagavad-Gita: Krishna's Counsel in Time of War*, trans. Barbara Stoler Miller (New York: Bantam, 1986), 27.

7. *Bhagavad Gita: A New Translation*, trans. Stephen Mitchell (New York: Three Rivers, 2000), 47.

8. Miller, tr., *The Bhagavad-Gita: Krishna's Counsel in Time of War*, 33.

9. Ibid., 99–100.

10. *The Yoga Sutras of Patanjali: A New Edition*, trans. and commentary, Edwin F. Bryant (New York: North Point, 2009).

11. Pandit Rajmani Tigunait, *The Secret of the Yoga Sutra: Samadhi Pada* (Honesdale, PA: Himalayan Institute, 2018), xiii.

12. Ibid., xiv.

13. Ibid.

14. Ibid., xv.

15. Ibid., xvi.

16. Ibid., xix.

17. Bryant, tr., *The Yoga Sutras of Patanjali*.

18. Ibid., 301–10.

19. Brigid Schulte, "Harvard neuroscientist: Meditation not only reduces stress, here's how it changes your brain," *Washington Post,* May 26, 2015. https://www.washingtonpost.com/news/inspired-life/wp/2015/05/26/harvard-neuroscientist-meditation-not-only-reduces-stress-it-literally-changes-your-brain/

20. Ibid.

21. Chittaranjan Andrade and Rajiv Radhakrishnan, "Prayer and healing: A medical and scientific perspective on randomized controlled trials," *Indian Journal of Psychiatry,* Vol. 51, no. 4 (October–December 2009), 247–53. https://www.ncbi.nlm.nih.gov/pmc/articles/PMC2802370/

22. Ibid.

23. Ibid.

24. Dean Radin, *Entangled Minds: Extrasensory Experiences in a Quantum Reality* (New York: Pocket Books, 2006), 198–207.

25. Lao Tzu, *Tao Te Ching*, trans. Stephen Mitchell (New York: Harper, 1998), from the Foreword.

26. Ibid.

27. Ibid.

28. Robert Stickgold, "Sleep on It!" *Scientific American*, October 2015.

29. Ibid.

30. Ibid.

31. Jack Hass, *The Dream of Being: Aphorisms, Ideograms, and Aislings* (Vancouver, BC: Iconoclast Press, 1966), 5

32. Tigunait, *The Secret of the Yoga Sutra*, 109.

33. Ibid., 124.

34. Bryant, tr., *The Yoga Sutras of Patanjali*, 345.

35. Ibid., 352.

36. Ibid., 353.

37. Ibid., 356.

38. Ibid., 357.

39. Ibid., 369.

40. Ibid., 375.

41. Ibid., 377.

42. Ibid., 378.

43. Ibid., 368.

44. Tigunait, *The Secret of the Yoga Sutra*, 178.

45. Bryant, tr., *The Yoga Sutras of Patanjali*, 368.

46. Tigunait, *The Secret of the Yoga Sutra,* 178–79.

47. James Allen, *James Allen 21 Books: Complete Premium Collection* (Scotts Valley, CA: CreateSpace, 2016), 1.

48. Ibid., 2.

49. Joseph Murphy, *The Power of Your Subconscious Mind* (Mansfield Centre, CT: Martino, 2011; first published 1963 by Prentice-Hall), 37.

50. Ibid.

51. Mitch Horowitz, "The Substance of Things Hoped For: Searching for Neville Goddard," introduction, Neville Goddard, *The Neville Reader* (Camarillo, CA: DeVorss, 2005), xi.

52. Ibid., xiii.

53. Ibid.

54. Ibid., xiii.

CPSIA information can be obtained
at www.ICGtesting.com
Printed in the USA
BVHW041344100422
633498BV00003B/12